Optimization in industry

Optimization in industry

VOLUME TWO
Industrial applications

T. A. J. Nicholson

Routledge
Taylor & Francis Group

LONDON AND NEW YORK

First published 2008 by Transaction Publishers

Published 2017 by Routledge
2 Park Square, Milton Park, Abingdon, Oxon OX14 4RN
711 Third Avenue, New York, NY 10017, USA

Routledge is an imprint of the Taylor & Francis Group, an informa business

Library of Congress Catalog Number: 2007019719

Library of Congress Cataloging-in-Publication Data

Nicholson, T.A.J.
    Optimization in industry / T.A.J. Nicholson.
        p. cm.
    Originally published: Chicago : Aldine Pub. Co., 1974.
    Includes index.
    ISBN 978-0-202-30965-1 (acid-free paper)
    1. Programming (Mathematics) 2. Mathematical optimization. I. Title.

T57.7.N53   2007
658.4'033—dc22—dc22   2007019719

ISBN 13: 978-0-202-36156-7 (pbk)

*For Claire
Graham and Diana
and all other
optimists*

Acknowledgements

I am grateful to Professor Moore of the London Business School for his frequent encouragement in the course of writing this book. The radical changes and improvements in the text since its first draft are largely due to the students of the first two M.Sc. courses held at the Business School and members of staff, particularly to Stewart Hodges: I am extremely grateful to them and to all other critics although my appreciation was rather less at the time of the suggested alterations. My thanks are very much due to my mother for meticulously inserting all the corrections in an already messy typescript. I could not face the proof reading alone, and I must thank Roger Garside, formerly of the A.E.R.E., Harwell, for undertaking the bulk of the work at that stage.

Contents

Foreword

This series arises directly out of the experience of teachers concerned with a range of (post)-graduate and post-experience courses, from Masters and Ph.D programmes, taught primarily to recent graduates of extremely high calibre but from very diverse disciplines, to programmes for executives, often graduates of some years' standing, in mid-career in management. In developing all of our courses from scratch (the school opened in 1965) we quickly found that while some of the available textbooks were clearly the product of a basic philosophy of teaching akin to our own, they were yet in general inadequate for our purposes. We held that education for management is essentially creating the ability to analyse business problems to formulate viable solutions in a changing environment and to develop skills in bringing innovation about. It requires a comprehensive programme of studies depending on mathematical competence, financial knowledge, marketing expertise, a comprehension of the economic, technological and social forces working upon business, and a systematic study of individual and group behaviour. We found that the shortcoming of existing texts in following these aims was the lack of a sufficiently broad and rigorous, yet *consistent* coverage of materials. Some areas of teaching are of course, better served than others, and this largely accounts for the sequence in which titles for the series are prepared— the most urgent needs, as we saw them, being met first. Thus we aimed on the one hand to produce texts suitable for adoption in the quantitative areas of a two-year Masters' programme and on the other, high-level and intellectually satisfying expository texts, again in the quantitative areas, suitable for adoption in post-experience programmes.

Partly also, our experience, and thus our perception of the functions textbooks should serve, has been formed by a critical but exceptionally constructive student audience. Many University teaching institutions are going through what is often felt to be a traumatic period of rising, and more articulate customers' demand for clear objectives in courses and relevant and economically presented material. As a new institution we have lived with the need continually to query and adjust objectives and methods right from the beginning. Students can no longer be called an 'audience', rather they are active participants in the teaching process. This is reflected in the textbooks in the series in that each results from practical classroom trials; participants on courses have contributed significantly to the final shape the texts take.

The teaching task in a graduate business school has, however, been shown to be inseparable from other functions an academic performs—notably consultancy and research. Good teaching depends on the stimuli of a constant challenge to make the material taught capable of application, and of contributing to the development of knowledge in the basic disciplines. An active professional life, in both these senses, complements the teaching work in such a school. Much of the writing arising out of such activities of course finds its way into the traditional avenues of publication. But some is more appropriately published in book form. The series thus will cover research monographs, collections of papers and proceedings of conferences of immediate interest to graduates in industry, and case materials.

M. E. BEESLEY

Preface

Volume II is a detailed review of the application of optimization techniques in industrial practice. It begins with a discussion of how to create mathematical models of industrial functions, such as process control, scheduling, design, replacement and distribution. The examples and case studies are taken from a wide range of industrial contexts and each application is formulated in detail up to the point where it can be solved directly by one of the techniques described in volume I.

1 Industrial models for optimization

1.1 The construction of models

The origin of any industrial optimization study lies in a hunch that some improvement can be made in a controllable system: the possibility may arise in any context: for example in the control of a chemical plant, in the organization of production to meet delivery dates, in the design of rubber compounds, in traffic signal settings, and so on. The first task towards solving the problem is to assess all the facts about a problem and consider their interactions. This is called the construction of a model.

The popular image of a model of an industrial process is a working miniature of the real thing. Here we will not be concerned with mechanical models, but with mathematical models which define and relate all the factors in a problem. This is the first stage towards formulating an optimization problem. If we wish to route a delivery vehicle through a number of distribution points we cannot physically set out with a vehicle to try the possible routes. Instead we build a model which can be used to calculate the effect of choosing different routes. If we wish to study what kind of reservoirs should be developed for the water supply system we cannot physically try out various places in future years; instead we propose a model of the supply system and impose the development policies on the model to see what will be the effects. Model construction is a skilful art and it can be one of the most fascinating exercises in optimization studies.

Models are used in such a wide range of contexts and for such differing purposes that it is impossible to give a single all-embracing definition of how a general model is made up. However, in the following sections we will attempt to outline the way in which one proceeds.

1

In the industrial field a model begins with a description of a physical system. This often includes a mass of information: lists of components of the system such as machines, processes, and storage points, their physical specification such as measurements of capacities and processing rates, and the information on technical restrictions and costs. The first step towards specifying the model is to establish from this information all the inter-relationships between the factors It is then often necessary to simplify the information and limit the scope of the model. This is achieved by making a set of assumptions about the system so that it takes a workable shape. The next task is to specify the relationships between the factors of the system in mathematical form. These relationships are often obtained by regression techniques: Section 1.4 is devoted to an outline of the least squares regression procedure.

When the relationships which define the model have been specified there are two further factors which may give rise to difficulty in the use of the model. First the relationships may not be permanent but change over time; in this case it is necessary to develop an adaptive model which adapts to the changes. (The adaptation scheme is discussed in Section 1.5.) Secondly the model may be too large to be workable as the subject of optimization. In this case it may be possible to partition the problems into subsystems which can be optimized one at a time. (The isolation of subsystems is discussed in Section 1.6.)

The final two sections consider the types of optimization policies which may be applied in industrial processes.

1.2 Assumptions, approximations, estimations

The purpose of a model is to provide a copy of a real system. We need to be able to use the model to predict the future outcomes of actions taken to control the real system. Very accurate models can sometimes be achieved. The amazing accuracy of space probes and moon flights is a consequence of nearly perfect models of gravitational fields and their interactions. Industrial models can rarely attain this level of perfection. It is not simply a question of inadequate knowledge about the system being studied; it also comes from a need to simplify the system and reduce the problem to manageable propor-tions for mathematical treatment. The more complex the model the more difficult will it be to conduct the optimization calculations and the longer will be the computing time required. We should make as many simplifications to the model as possible as long as the answers remain realistic and can be used for the purposes for which they are intended.

There are three types of simplifications which are made in one form or another to nearly all models. We will distinguish them as assumptions, approximations and estimations, although to some extent they all overlap.

The assumptions define the limitations of the model and the structural simplifications which have been made. Above all, the assumptions state which considerations have been left out of the model either deliberately or through lack of knowledge. The aim of an assumption is to gain in model simplicity without too much loss of realism.

The approximations are mathematical decisions to approximate a complex function by a simplified function. A complex relationship between one factor and a number of other factors may be approximated by a linear or quadratic relationship by using regression techniques described in the section 1.4. The aim of an approximation is to gain in mathematical simplicity without too much loss in accuracy.

The estimations are statistical decisions assigning values to parameters of the model which depend on some stochastic or random element which cannot be controlled. For example, in the allocation of water from reservoirs we may use an estimate that monthly rainfall levels will be equal to the average monthly rainfall over the previous five years. The aim of using estimates is to gain in the deterministic nature of the variables without invalidating the answers through ignoring an important random element.

Example 1.1

The annual yield of agricultural land is assumed to depend on the number of hours of sunlight and the quantity of fertilizer applied. This is an assumption as the yield is in fact influenced by other factors such as rainfall, the distribution of rainfall over the year, what was grown the previous year and so on. But we probably do not understand the interactions fully, and so we have to make this simplification. We would write the assumed relationship that the value of the yield y is related to sunlight s and fertilizer level x as

$$y = as + bx$$

where a and b are constants. This is, in effect, an approximation to the real relationship

$$y = f(s, x)$$

as s and x probably interact creating a non-linear function.

Sunlight is of course not known accurately and we can only estimate what the value will be by recording sunlight from previous years together with meteorological predictions. So we have to make estimates in this problem as well.

Example 1.2

A distributor wishes to determine the optimal policy for routing his vehicles. It is assumed that the daily carrying capacity of a truck is constant despite the variation in traffic conditions, causing delays occur at different times of the day and in different geographical areas. The routing may be planned by approximating the actual mileage between two points as proportional to the straight line distance. If $m(i, j)$ denotes the actual mileage and $d(i, j)$ denotes the straight line distance we approximate $m(i, j)$ as

$$m(i, j) = k \cdot d(i, j)$$

where k is a constant of proportionality. We might use different values of k for urban and rural areas. The demand points may be defined as a collection of individual customers all occurring in a local area and this demand group is approximated by a point on the map with the delivery requirements added up. It is an approximation because we know what the exact demand pattern is in detail but want to simplify the problem. The planned routings for the next week may be based on a weekly adjusted estimate of demand for the previous three weeks at each point, thus eliminating the random element.

1.3 Assessing interrelationships

When the scope of the model has been defined the interrelationships between the factors in the problem must be assessed and expressed mathematically. This is a crucial step: the relationships form the constraint equations and inequalities of the optimization problem. It could be disastrous if one binding interrelationship remained unobserved.

Whatever the context of the problem, the constraints will typically occur as equality or inequality conditions on the problem variables or commitments for certain variables to be discrete valued. For example, in chemical process control the system often requires a mass balance equation to hold. If three materials A and B and C are entering a reactor with masses per unit time m_a, m_b, m_c and two products D and E are emerging as the yield with masses m_d and m_e per unit time, the mass balance equation required that

$$m_a + m_b + m_c = m_d + m_e.$$

Inequality relationships can arise where it is necessary to process a product in a definite sequence. If $x(j, k)$ denotes the start time of job j on process k and the duration of the process is $d(j, k)$ we will require that the start time of the next process $x(j, k+1)$ is greater than the finish time of the job on the previous process, i.e.

$$x(j, k+1) \geqq x(j, k) + d(j, k).$$

Similarly inequality constraints occur as limits on aggregates. Suppose an advertising campaign is being planned and the cost of inserting an advertisement in media i is c_i per insertion of the advertisement. If the total amount which can be spent on the campaign over N media is C and x_i is the number of insertions in media i we require that

$$c_1 x_1 + c_2 x_2 + \ldots + c_N x_N \leqq C.$$

Controls on machinery or limits on property requirements of materials often require inequalities of the form

$$a \leqq x \leqq b.$$

Also there are a lot of problems where variables must be integer valued—the available diameters of a pipe or the number of vehicles in a queue. The type of constraints which we obtain will decide the optimization techniques which are available for solving the problem, and we should be aware of this as the model is being developed so that appeoximations can be made where appropriate.

1.4 Proposing mathematical relationships: regression techniques

Many of the relationships between factors may not be known as mathematical forms. We may only suspect that a relationship exists between them. For instance, it may be believed that the value of one variable depends on the values of certain other variables which we can control. But we may not know the exact form of the dependence. All we can do is to propose a trial mathematical relationship between the dependent variable and the control variables and then try out the relationship experimentally to see if it is reasonable. There is no source of advice on how to propose relationships—we should clearly choose as simple an expression as possible such as a linear or quadratic form. However there are established procedures for testing out the adequacy of proposed relationships; these methods are known as regression techniques and we will describe these in detail.

The object of regression techniques is to choose a 'best' mathematical relationship out of a class of possible relationships on the evidence of some experimental data. Suppose we proposed an exact function for the relation between two quantities y and x (where x is the controllable variable) as

$$y = 3 + 0.24x.$$

We are stating that the value of y is equal to 3 plus a component proportional to the value of the variable x. For example we may be measuring agricultural yield as a function of the amount of irrigation. Now we may be correct in the assumption that y is the sum of a constant term and a factor proportional to x, but it is highly improbable that, without any other knowledge, we can accurately guess the values 3 and 0.24 in the relationship. A much wiser plan would be to propose that the function has the looser but more general form

$$y = a_0 + a_1 x$$

where a_0, a_1 are not initially specified. a_0 and a_1 are called the parameters of the relationship. The problem would now be to choose the most suitable values for a_0 and a_1. Some criterion is needed for what we mean by suitable. Clearly we wish to minimize the discrepancies between the model and the observed facts. We will therefore take a number of trial values of x around the region of interest, measure the corresponding y values and choose the a_0 and a_1 quantities to minimize the differences between the observations and the model predictions.

Suppose we were interested in the value of x in the range 12 to 48. We might take seven x values at $x = 12, 18, 24, 30, 16, 42$ and 48 and record them as $x^{(k)}$, $k = 1, ..., 7$. The y values corresponding to these x values might be as shown in Table 1.1. The points $(x^{(k)}, y^{(k)})$ are plotted in Fig. 1.1 together with a line of the form $a_0 + a_1 x$. Now this does not pass through all the points, and indeed no straight line could go through all points, as the points $(x^{(k)}, y^{(k)})$ are not all in a straight line. The difference between the actual $y^{(k)}$ values and the values predicted by the line, say $\hat{y}^{(k)}$, are marked in the diagram

Table 1.1

k	1	2	3	4	5	6	7
$x^{(k)}$	12	18	24	30	36	42	48
$y^{(k)}$	5·27	5·68	6·25	7·21	8·02	8·71	8·42

as the deviations $d^{(1)}$, $d^{(2)}$, $d^{(3)}$, $d^{(4)}$, $d^{(5)}$, $d^{(6)}$, $d^{(7)}$. We wish to choose the a_0 and a_1 values to minimize the deviations. This is an optimization problem in

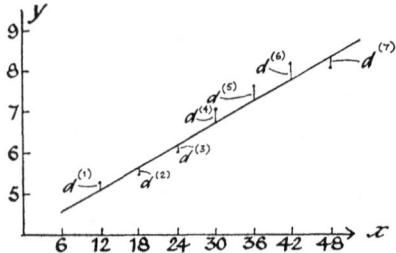

Fig. 1.1

its own right. In fact the standard regression procedure is to choose the values of a_0 and a_1 to minimize the sum of the squared values of $d^{(k)}$, as it is analytically simpler and it stresses the avoidance of large deviations. We therefore minimize:

$$S = (d^{(1)})^2 + (d^{(2)})^2 + (d^{(3)})^2 + (d^{(4)})^2 + (d^{(5)})^2 + (d^{(6)})^2 + (d^{(7)})^2.$$

As $d^{(k)} = y^{(k)} - (a_0 + a_1 x^{(k)})$ for $k = 1, ..., 7$, the function we wish to minimize is

$$S(a_0, a_1) = \sum_{k=1}^{7} (y^{(k)} - a_0 - a_1 x^{(k)})^2.$$

The only unknowns in this function are a_0 and a_1. We can therefore differentiate this function with respect to a_0 and a_1 to give

$$\frac{\partial S}{\partial a_0} = -2 \sum_{k=1}^{7} (y^{(k)} - a_0 - a_1 x^{(k)})$$

$$\frac{\partial S}{\partial a_1} = -2 \sum_{k=1}^{7} x^{(k)}(y^{(k)} - a_0 - a_1 x^{(k)}).$$

Equating these two derivatives to zero for a minimum we get

$$\sum_{k=1}^{7} y^{(k)} = 7a_0 + a_1 \sum_{k=1}^{7} x^{(k)}$$

and

$$\sum_{k=1}^{7} x^{(k)} y^{(k)} = a_0 \sum_{k=1}^{7} x^{(k)} + a_1 \sum_{k=1}^{7} (x^{(k)})^2.$$

If we now substitute in these equations using the values of $x^{(k)}$ and $y^{(k)}$ of Table 12.1 we get two equations in a_0 and a_1 as

$49{\cdot}56 = 7a_0 + 210a_1$

$1580{\cdot}5 = 210a_0 + 7308a_1.$

These can be solved for a_0 and a_1 to give:

$a_0 = 4{\cdot}0,\ a_1 = 0{\cdot}1.$

The results are often called estimates of a_0 and a_1 and are denoted by $\hat{a}_0$, $\hat{a}_1$. The required relationship is thus

$y = 4{\cdot}0 + 0{\cdot}1x.$

This completes the regression method for determining a_0 and a_1.

We will now illustrate regression on a more general case as the ideas are very important in model construction. Suppose we had one factor y which was related to N other factors $x_1, x_2, \ldots x_N$. The quantity y might be the plasticity of a chemical product and the quantities x_1, x_2, x_3, x_4 the proportions of four constituents which are necessary to make the product. If the independent factors x_i do not interact with one another a linear relationship between the property y and the constituents might be proposed such as

$y = a_0 + a_1 x_1 + a_2 x_2 + \ldots + a_N x_N$

where $a_1, a_2, \ldots, a_N$ are the unknown constants of proportionality. However if there were some interactions between the factors, it may then be more appropriate to propose a quadratic relationship such as

$$y = a_0 + \sum_{i=1}^{N} a_i x_i + \sum_{i=1}^{N} \sum_{\substack{j=1 \\ j \geq i}}^{N} a_{ij} x_i x_j.$$

We will presume that the linear relationship is appropriate for this case.

To obtain values for the coefficients a_i we perform, as before, a number of trial experiments for different values of $x_1, x_2, \ldots, x_N$, and record the values of y for each case. The question of which points x_i to choose would be more difficult. The choice properly belongs to the subject of experimental design and we will not discuss it further here. We will clearly choose points in the region of interest for the optimization study so that the relationships are accurate where the optimization search takes place. If there are M trial points, the values of y and $x_1, x_2, \ldots, x_N$ will be recorded at each point and may be denoted by $y^{(k)}$, $x_1^{(k)}$, $x_2^{(k)}$, $x_N^{(k)}$ for $k = 1, ..., M$. We now wish to 'fit' the relation $y = a_0 + a_1 x_1 + ... + a_N x_N$ to these values. If there were just $N+1$ sets of values we could write this relation for the values $y^{(k)}$ and $x_j^{(k)}$ and solve the resulting equations for the parameters $a_0, a_1, ..., a_N$. Usually it is desirable to have more sets of observations than parameters and we can again adopt the regression policy of determining the parameters to minimize the sum of the squared deviations of the observed measurements from the model or theoretical values which would be predicted by the relationship. The theoretical values of y say $\hat{y}^{(k)}$ are given by the equation

$$\hat{y}^{(k)} = a_0 + a_1 x_1^{(k)} + a_2 x_2^{(k)} + ... + a_N x_N^{(k)}$$

so that the sum of the squared deviations given any values of $a_0, a_1, ..., a_N$ is

$$S(a_0, a_1, ... a_N) = \sum_{k=1}^{M} (y^{(k)} - \hat{y}^{(k)})^2$$

$$= \sum_{k=1}^{M} [y^{(k)} - (a_0 + a_1 x_1^{(k)} + a_2 x_2^{(k)} + ... + a_N x_N^{(k)})]^2.$$

We can now minimize S with respect to the parameters a_i by differentiating and equating the resulting derivatives to zero. The equations are:

$$\frac{\partial S}{\partial a_i} = -2 \sum_{k=1}^{M} x_i^{(k)} [y^{(k)} - (a_0 + a_1 x_1^{(k)} + ... + a_N x_N^{(k)})] = 0 \text{ for } i = 0, 1, 2, ... N$$

and $x_0^{(k)} = 1$ for $k = 1$ to M.

Solving these equations we obtain estimates for a_i as $\hat{a}_i$. The values of $\hat{a}_i$ may be neatly expressed in matrix notation as

$$\hat{A} = (X'X)^{-1} X'Y$$

where $\hat{A}$ is the column vector of $\hat{a}_i$ values for $i = 0, ..., N$, Y is the column vector of $y^{(k)}$ values, $k = 1, ..., M$, and X is the $Mx \times (N+1)$ matrix of $x_j^{(k)}$ values

$$X = \begin{pmatrix} 1 & x_1^{(1)} & x_2^{(1)} & ... & x_N^{(1)} \\ 1 & x_1^{(2)} & x_2^{(2)} & ... & x_N^{(2)} \\ \vdots & \vdots & \vdots & & \vdots \\ 1 & x_1^{(M)} & x_2^{(M)} & ... & x_N^{(M)} \end{pmatrix}.$$

We have thus obtained the required relationship between y and the quantities $x_1, x_2, ..., x_N$. If a more complex quadratic relationship had been presumed the procedure would have been identical as the function is still linear in the coefficients a_i and a_{ij}. If the proposed function relating the factors is non-linear in the coefficients we encounter the usual difficulties associated with minimizing non-linear functions.

Example 1.3

The properties of rubber washers, belts and hoses are assumed to depend linearly on the proportions of the ingredients which enter the initial mix for the product. The products are manufactured by mixing a percentage of raw rubber, parts of oil, parts of crumb (scrap) and a percentage of carbon black. The properties of a particular class of items being studied are tensile strength and break length. Eight sets of values of the variables were taken and the property levels measured. The results are shown in Table 1.2

Table 1.2

Observation k	Raw rubber $x_1^{(k)}$	Oil $x_2^{(k)}$	Crumb $x_3^{(k)}$	Carbon black $x_4^{(k)}$	Tensile strength $y_1^{(k)}$	Break length $y_2^{(k)}$
1	65	10	85	15	890	310
2	35	30	85	15	1140	350
3	65	30	85	85	1410	370
4	35	10	15	15	1240	430
5	35	30	15	85	1560	410
6	35	10	85	85	1650	370
7	65	10	15	85	1340	390
8	65	30	15	15	950	420

A linear relationship was assumed between the properties (denoted by y_1 and y_2) and the variable values, and the least squares method was used to obtain the resulting equations as

$$y_1 = 1393 \cdot 5 - 8 \cdot 333 x_1 - 0 \cdot 750 x_2 - 0 \cdot 00002 x_3 + 6 \cdot 214 x_4$$

$$y_2 = 437 \cdot 2 - 0 \cdot 583 x_1 - 0 \cdot 625 x_2 - 0 \cdot 893 x_3 + 0 \cdot 107 x_4.$$

It may be observed that in the first equation the coefficient of x_3 in the equation for y_1 is very small. This suggests that it may be adequate to assume that the relationship for y_1 can exclude the component for x_3. This is a statistical question and the schemes for reducing proposed relationships to their minimal size are discussed in texts on statistics.

1.5 Adaptive models

If mistakes are made in the construction of a model this may not be a disaster. We may have opportunities to revise the model and improve the accuracy of relationships which have been proposed. Indeed we may deliberately over-simplify some characteristics of the system and plan definite times for revision. The adaptation may be undertaken automatically by a computer. These self-adaptive systems are particularly relevant in dynamic situations such as those occurring in chemical process control.

Suppose a complex system is defined in terms of a set of relationships between a number of dependent and independent variables. If the system is changing over time then the model will have to be altered over time for control purposes. The rate at which we have to recalculate optima may demand that we establish comparatively simple mathematical forms for the model so that the computations are quick. Starting at some given time we would propose functions for the interrelationships and for the objective function based on the data we had available and we would use these functions to calculate an optimum. We would then implement the control settings which we have just determined in the model on the real system. This may mean reading the output from a computer and adjusting the values on some valves in a chemical plant or automatically altering a pattern of signal settings in a traffic control scheme. After implementing the optimum we may find that the nature of the objective function and constraints are not exactly what we thought they would be at the optimum and it is therefore worth revising the coefficients in the objective

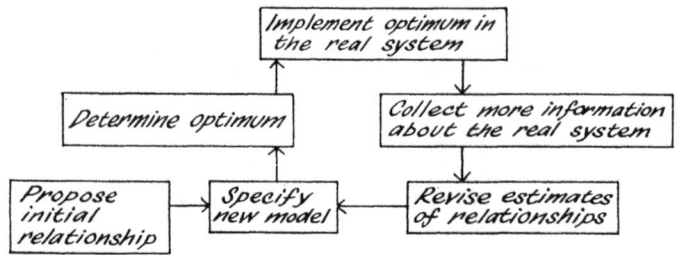

Fig. 1.2

and constraint functions on the basis of this new information. Indeed, we may recalculate a regression simply on the new evidence, but normally we would base it on past and present information. Once the new relationships have been determined a new optimum can be calculated and implemented, and again the relationships are revised as more evidence of the system becomes available.

The revision scheme is not only necessary for systems which change rapidly over time. It is also relevant for large complex systems when we may simply be learning about the system and gradually understanding how to control it as we adapt the model to specify it more exactly.

The scheme of proposing a model, determining the optimum, implementing the result and revising the model is displayed in the flow chart of Fig. 1.2. It is important to realize that although this adaptive scheme is well suited to dynamically changing systems, the system must be stable enough or the calculations fast enough to be able to determine a new optimum while the relationships still provide a reasonable representation of the problem.

1.6 Isolating subsystems

Very large systems may be intractable because of their sheer size. If there are hundreds or even thousands of variables which we wish to control there may be no optimization techniques which can handle the problem. Often, however, large systems can be broken down into subsystems which can be optimized separately. The method of dividing up the problem may be almost self-evident. The separation may be made by space or geographical considerations or by time intervals. Other possible criteria for dividing up the problem could be by long- and short-term factors, or by separating the various technical aspects. Three examples of this sort of isolation of a large problem into subsystems follow.

Example 1.4

The total problem of producing petrol includes numerous stages. It begins with the extraction of crude oil from oil wells. This is transported to refineries by oil tankers where it is refined and blended into various grades of aviation fuel and petrol. Subsequently it is held in storage tanks from which it is distributed to suppliers.

Fig. 1.3 illustrates the stages in the total process. The total problem should

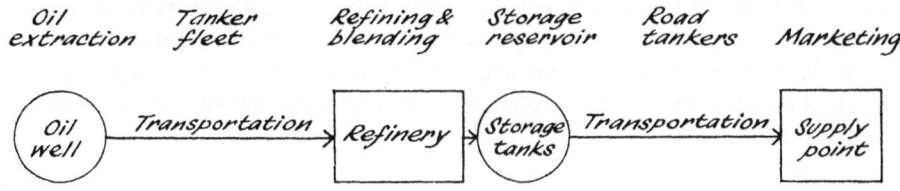

Fig. 1.3

include the search for crude oil, the size and organization of oil tankers, which grades of petroleum to manufacture, the scale of the reserve stocks and the location of supply points and the marketing policy to customers. The decisions on all these factors affect the overall efficiency of the system, and all the parts interact to some extent. But clearly the individual stages can be treated as independent subsystems. They comprise the oil extraction, the transportation to refineries, the refining and blending, the storage and final distribution systems, and each subsystem may be optimized separately.

The means of relating the results of one subsystem optimization to another will vary. The simplest way of resolving the connections is to solve one problem after another in a sequential fashion. Suppose we have three subsystems with problem variables as:

Subsystem 1: $X = (x_1, x_2, ..., x_M)$

Subsystem 2: $Y = (y_1, y_2, ..., y_N)$

Subsystem 3: $Z = (z_1, z_2, ..., z_P)$.

We may be able to solve subsystem 1 and determine the X values, feed these into subsystem 2 as fixed quantities, determine Y and then feed X and Y into subsystem 3 to determine Z. The petrol production process of Example 1.4 could perhaps be treated in this way. But the subsystems are rarely sufficiently independent of one another to make this sequential treatment valid. We may have to revise the solution to subsystem 1 on a result of the solution to subsystem 3, and then revise subsystem 2 and so on. Solving subsystems cyclically in this way has occasionally proved useful, but unless the information about each subsystem is steadily improving as the cycles proceed the approach could prove unstable and unsuccessful. Yet another possibility is to select certain variables as the link variables between the subsystems and allow them to be variables in each subsystem, but it may be difficult to find a few suitable variables.

Very few general conclusions have been reached on the subject of how to isolate subsystems and connect up their individual results. The only computational development in this area is the decomposition technique in linear programming which enables large sparse matrices to be treated as a series of smaller linear programming problems. The judgements about how to partition a large system into subsystems and how to take account of the interactions will often have to be intuitively based on the structure of the particular application. Often it will be difficult to know if the best type of partitioning has been achieved. The important point is to realize that an optimization study is nearly always treating one aspect of a larger system, and what appear to be rigid constraints in one problem may be free variables in another. Often this passes unnoticed.

1·7 Optimization policies

A model of a system is not itself an optimization problem: it is a definition of an industrial system. The optimization problem arises when a selected set of factors or quantities in the model are chosen as the control variables and an objective is defined. After the control variables have been picked out, all the remaining quantities and symbols in the list of notation are presumed to be available as data. The relationships which have been derived to describe the behaviour of the model are now assembled as constraints on the problem variables and the system objective is written down as the objective function. Only at this point do we call upon optimization techniques: we use the techniques to find the settings for the control variables which will enforce the system to behave in accordance with the objective within the limits imposed by the constraints.

The choice of problem variables is obvious in some cases where it is clear what it is that we wish to optimize. But most industrial models contain a number of optimization problems, and if we look at the overall system from the point of view of minimizing costs we may come up with a variety of alternative factors to control. We might see the task of expanding production of a chemical as being a need for a larger plant. But equally we may be able to optimize the controls on the existing plant and increase production adequately in this way. We may view a congestion problem from the point of view of improving control of the items being congested within the given resources or expanding the resources to reduce congestion to a given level. Similarly in the optimal use of a bus fleet over an urban area, the model of the system will probably contain two sorts of problems, the fleet size problem and the scheduling problem. The scheduling problem is: Given a fixed fleet of vehicles, what schedule of arrivals and departures will give the highest level of service? The fleet size problem is: If a schedule of arrivals and departures is given, what is the minimum fleet size which can accomplish it? In these cases, one factor provides a set of problem variables and the other acts as a constraint and it is important to be aware of the alternative optimization problems which can be embedded in any one system.

Generally we shall encounter three types of objectives in industrial contexts: the minimization of costs, the maximization of profits or the minimization of time taken to complete a given activity. In stock holding and production scheduling we would minimize costs and for plant renewal we might minimize discounted costs over a period of time. In investment schemes or the choice of products which should be produced the objective will often be to maximize profits. The third kind of objective, of minimizing an elapsed time, occurs in short-term problems such as machine sequencing and batch chemical processes where we wish to complete a given load of work or convert a given batch of material in the shortest time.

1.8 Applications of optimization in industry

In all the future discussion we will be formulating models, identifying controls, specifying objectives and interpreting the model relationships as constraints on the controls. The use of models will become a habit. But it is worth

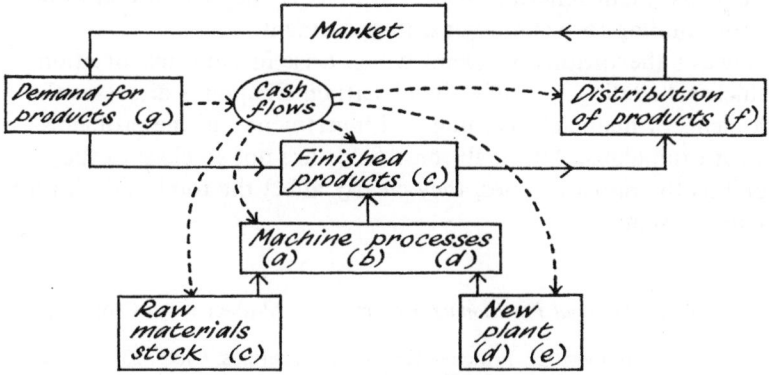

Fig. 1.4

collecting up the range of applications of optimization techniques at this point to see how the models which we shall be dealing with interact among themselves.

The model of Fig. 1.4 illustrates a general production system. The system begins with a market creating a demand for finished goods which are supplied through a distribution system. The stock of finished goods is supplied by a production unit consisting of machine processes which convert raw materials into finished goods. New products may be introduced into the system and the machines themselves are replaced by a new plant whenever technological improvement suggests this is advantageous. It should be noted that the return to the production system depends entirely on maintaining the demand for its products. The distribution of goods, stock holding of finished goods, purchase and storing of raw materials, the operation of machines and the introduction of new plant are all costs. These are indicated in Fig. 1.4 by the dotted lines showing the cash out-flows and the single cash in-flow.

The first two application areas which we will study in Chapter 2 and 3 arise in the production unit: process control of chemical plant and the sequencing of jobs through machines to meet delivery dates. Both these problems are short-term, primarily being concerned with the efficient operation of equipment and the utilization of resources.

The next subject, stock control and production scheduling studied in Chapter 4, applies to the efficient management of stocks either as raw material or as finished products and the setting of production targets to meet demand and smooth the use of resources. These are the medium-term problems of the production unit which can take account of the market and fluctuations in prices and costs.

Product design and plant renewal problems which are studied in Chapters 5 and 6 are both long-term issues. Design optimization problems arise in the choice of raw materials to manufacture products to meet certain properties, or the choice of technical specifications of a machine to satisfy certain requirements. Plant renewal problems consider the task of renewing machinery corresponding to technological improvement.

Chapter 7 reviews the various problems which arise in the transportation field from vehicle scheduling to traffic signal setting. Chapter 8 discusses the long-term problems of financial planning and investment and the problems of advertisement campaigns. Although these fields are not so close to the production unit as the previous ones, they clearly affect the total optimization of the production system.

REFERENCES

Anthonisse, J. M. 1968. *An Optimal Partitioning.* Report S398. Mathematisch Centrum, Amsterdam.
Dantzig, G. B. 1963. A Decomposition Principle for Linear Programs. Chap. 23 of *Linear Programming and Extensions.* University Press, Princeton.
Eckman, D. P. and Lefkowitz, I. 1960. *Principles of Model Techniques in Optimizing Control. Proc. First Congress Int. Fed. Autom. Control Moscow.* Butterworth, London.
Harling, J. 1958. Simulation techniques in operational research. *Opl Res. Quat.,* 9 (1).
Johnson, R. A., Kast, F. E. and Rosenzweig, J. E. 1964. Systems theory and management. *Man. Sci.* 10 (2), 367.
Morgenthaler, G. W. 1961. The Theory and Application of Simulation in Operations Research. Chap. 9 in *Progress in Operations Research,* Vol. 1. Ed. R. L. Ackoff. J. Wiley, New York.

Morris, W. T. 1967. On the art of modelling. *Man. Sci.*, **13** (12).
Reisman, A. and Buffa, E. S. 1964. A general model for production and operations systems. *Man. Sci.*, **11** (1), 64.
Theil, H. 1965. Econometrics and management science: their overlap and interation. *Man. Sci.*, **11** (8), B200.
Wismer, D. A. 1967. On the uses of industrial dynamic models. *Opns Res.*, **15** (4).

Exercises on Chapter 1

1 A dependent variable y is assumed to be related to another variable x by the quadratic function

$$y = ax^2.$$

Four trial values of x are taken at the values $x = 0, 1, 2, 3$ and the measured values of y obtained are 0, 2, 2, 5 respectively. Determine the value of a by a least squares regression technique and note the greatest deviation between the actual and fitted values of y.

2 A proposed relationship between the factor y and variables x_1, x_2 is

$$y = ae^{-bx_1} + cx_2$$

where the coefficients a, b and c are to be determined. If N values $x_1^{(i)}$, $x_2^{(i)}$ are taken and the corresponding values $y^{(i)}$ are measured, describe how the least squares estimates of a, b and c could be determined.

*3 The criterion of minimizing the squares of the deviations between the actual and predicted values is not the only way in which proposed relationships between variables can be established. Suggest some other criteria. Why, for instance, should we not take the deviations unsquared, or the modulus of the deviation?

2 Chemical process control

2.1 Chemical process models

In this chapter we shall study the optimization problems arising in industrial processes which are in a fairly rapid state of change. Typically we shall be referring to processes in the chemical and petroleum industries, although the same character of optimization treatment may be applied to other dynamic problems such as those occurring in urban traffic control. The basic aim of

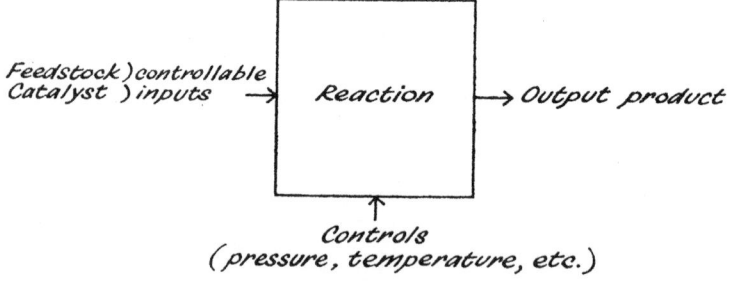

Fig. 2.1

the processes in the chemical industry is to convert a raw material into a more valuable product by a chemical reaction or refining procedure such as the conversion of crude oil into petroleum.

The general nature of a process in the chemical field consists of a number of inputs, referred to as the feed stock, a reaction between the inputs usually assisted by a catalyst, and an output of the desired product. The reaction may be controlled with regard to temperature, pressure, catalyst inlet rate and other factors. Fig. 2.1 illustrates the scheme. Some reactors treat a continuous flow of inputs and others react one batch of material at a time for a period called the holding time. The reaction itself is governed by chemical and physical laws and sophisticated mathematics may be necessary to describe the process.

Chemical process models are defined in terms of three types of variables: control variables, performance variables and disturbance variables. The control variables provide the means whereby the conditions of operation are adjusted such as flow rate, heat input, pressure, etc. The performance variables are the outputs, profit and any other criteria for the process. The disturbance variables are the environmental factors such as purity of feeds, prices, and ambient temperatures which influence the optimum operating conditions. Although the disturbance variables are fixed at any one time, it is their variation which makes the optimization of chemical processes an important and profitable proposition. We will denote the three types of variables at time t by the following notation:

control variables $X(t) = (x_1(t), x_2(t), ..., x_n(t))$

performance variables $Y(t) = (y_1(t), y_2(t), ..., y_m(t))$

disturbance variables $Z(t) = (z_1(t), z_2(t), ..., z_p(t))$

where the index t has been included to denote the time dependence. Sometimes the x_i variables are called the independent variables and the y_i are called the dependent variables.

The objective function in a chemical process is often to maximize profits or minimize costs over a given period of time say $(0, T)$. Profit will be a function of the control variables and the performance variables. Denoting the profit at time t by $F(X(t), Y(t))$, the required objective is to maximize an integral of the form:

$$\int_0^T F(X(t), Y(t))dt.$$

In other cases the objective of the process may not be expressed in terms of profits or costs: it may be too difficult to measure the profit because of vague economic factors. If the process is production limited rather than market limited so that all the material produced is sold, the maximization of production may be equivalent to the maximization of profit. Other possible criteria are the minimization of off-quality product, or the maximization of yield.

The constraints on the process may be of two sorts: restrictions on the controls, and the process equations. The restrictions on the controls are typically inequality constraints of the form

$$a_i \leq x_i(t) \leq b_i.$$

For instance, the diameter of a pipe will limit the maximum flow rate which can be attained and there may be a minimum flow rate for safety reasons. These constraints may be more complex when they occur as controls on the quality of the output or the conditions of the reaction. The inequality may then involve a function of the control variables of the form

$G(X(t)) \leq b.$

The process equations describe the reaction converting the input materials into the output. The development of these process equations requires knowledge of the physical and chemical laws governing the process. The law of conservation of mass and the first law of thermodynamics provide important sources of process relationships; material and energy balances are based on them. The relationships will often take the form of differential equations. Material balance equations may be written in terms of gross ingredients entering and leaving the process, in terms of distinct chemical compounds, or in terms of individual chemical elements. Where reactions are diffusion-controlled, classical diffusion theory may be employed. Heat transfer theory can be applied to kilns and metal-rolling furnaces. We will denote these process equations by functions of the form

$H(X(t),\ Y(t),\ Z(t)) = 0.$

We will now illustrate the development of process models and objectives on two examples. The first example considers a reaction in a plant to be in a steady state and it distinguishes the different types of variables. The second example illustrates the nature of the dynamic process equations.

Example 2.1

A catalytic reforming plant consists of one feed line producing two output streams of reformate and hydrogen-rich gas. A catalyst is also present in the reaction. Let F be the feed rate of the input in pounds per hour, and R, H be the hourly production rates in pounds of reformate and hydrogen-rich gas. Suppose a, b, c are the prices per pound of reformate, hydrogen-rich gas, and feed, and let C be the variable hourly cost which includes both a constant operating cost and a catalyst cost which depends on operating conditions such as ambient temperature. F is the independent control variable, R, and H, are performance variables and a, b, c and C are economic disturbance variables.

The objective is to maximize profit per hour less the hourly costs, i.e.

$P = aR + bH - cF - C.$

The control variable feed rate F may be limited to lie between zero and a maximum value m,

$0 \leq F \leq m.$

The process equations relate the variables H, R and C to F so that the objective function can be expressed in terms of the control variable. By a material balance:

$$F = R + H$$

the profit function can now be rewritten as

$$P = aR + b(F - R) - cF - C.$$

Also the variable cost C may be related to catalyst depreciation which depends on feed rate giving a function relationship

$$C = g(F),$$

and therefore

$$P = aR + b(F - R) - cF - g(F).$$

Finally if the reformate level R is related to F by a yield equation as

$$R = h(F)$$

the profit P is now expressed in terms of the single variable F as

$$P = ah(F) + b(F - h(F)) - cF - g(F)$$

and this may be maximized by any non-linear optimization technique which will take account of the constraint on F.

Example 2.2

A simple chemical system consists of a batch reactor containing a material y which decomposes after heating by a first order reaction to produce material z. Heat is to be applied so as to convert y into z up to a given extent. Let $y(0)$ be the initial concentration of y at time $t = 0$, and let $y(t)$ and $z(t)$ be the concentrations of y and z at time t. The control variable is the rate of heat input into the system which we will denote by $x(t)$. A typical objective for this reaction would be to determine minimum total heat input $x(t)$ over a given period of time $(0, T)$ which will achieve the required conversion of y into a given level of concentration of z. This objective would require the function $x(t)$ to be determined to

$$\text{minimize} \int_0^T x(t)dt$$

subject to $z(T) \geq m$

where m is the required concentration.

The quantities $z(t)$, $y(t)$ and $x(t)$ are related by a mass and heat balance. The rate of decrease in the concentration of y must balance the increase in concentration of z. This is expressed by the differential equation:

$$\frac{dy}{dt} = -\frac{dz}{dt}.$$

The solution to this equation enables $z(t)$ to be expressed in terms of $y(t)$.

B

Also as the reaction is 'first order' a typical expression for dy/dt may be the first order differential equation

$$\frac{dy}{dt} = -ke^{-a/h(t)} \cdot y$$

where $h(t)$ is the heat of the reaction at time t, and k and a are constants. The heat or energy balance requires that the rate of change of the reaction temperature may be accounted for by the difference between the current temperature $h(t)$ and the ambient temperature A, together with the heat input and the heat generated in the reaction say $H \cdot \dfrac{dy}{dt}$, to give another differential equation having the form

$$c_1 \frac{dh}{dt} = c_2(A - T) + x(t) - H \frac{dy}{dt}$$

where c_1, c_2 and H are constants. This heat balance equation enables us to determine $h(t)$ as a function of $\dfrac{dy}{dt}$ and $x(t)$, and hence, using the previous equations, we obtain $y(t)$ and $z(t)$ as functions of $x(t)$. Only at this point can we start thinking about how to determine $x(t)$ to minimize the integral objective function expressed above, as the constraint may now be expressed in terms of the $x(t)$ variable.

This example illustrates the complex form which the process equations may take when it is necessary to consider the dynamics of the reaction. Later we will see how dynamic programming can be used to solve these problems without explicitly needing to solve the differential equations in terms of classical differential equation theory.

The two examples distinguish two types of optimization problems which arise in chemical processes. The first type assumes the reaction to be in a steady state and the controls are manipulated in response to alterations in the environment as expressed by changes in the disturbance variables. The second type of optimization problem considers the way in which a reaction should be controlled in the course of the conversion process, assuming that the environment external to the reaction remains unchanged. A third type of problem arises when the process consists of a series of reactions and the separate stages have to be controlled collectively. We will examine these three kinds of problem in turn.

2.2 Steady-state optimization

A chemical reaction is sensitive to all the factors in its environment and the optimum control settings for a reaction will vary depending on how these factors change over time. We have already labelled these factors appropriately as disturbance variables, and most reactions will suffer such disturbances

quite frequently. It is easy to see how they occur. The chemical composition
of the inputs may vary as each new batch is supplied. For instance, crude oil
is a mixture of organic compounds extracted from different geographical
regions and the resultant crude oil will differ in its chemical composition.
Similarly, iron ores from different sources exhibit different characteristics, as
do different pulps for a paper mill. The ambient conditions such as temperature and humidity may vary. The equipment will age. The raw material costs
and the product prices will alter over time. All these disturbances will call
for different control actions needing different settings for the control variables
(temperatures, pressures, etc.), in order to achieve the desired outputs at the
correct quality standards. Therefore the general objective of maximizing
profit from the plant, can be achieved only by making adjustments in the
control variables from time to time.

The effect of a disturbance is shown in Fig. 2.2 for a hypothetical process
with a disturbance variable D and a control variable C. The contour lines of

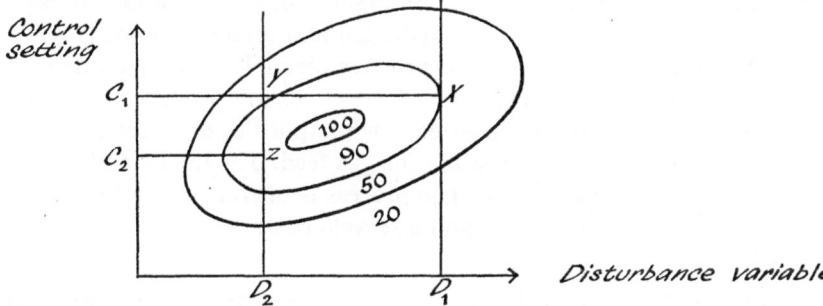

Fig. 2.2

the profit function are shown with the profit peak corresponding to 100 on
an arbitrary scale. At a particular time the disturbance variable value is D_1
and the control variable is set to maximize profit at C_1, the profit being the
height of the contour at X. If the disturbance variable then alters to the value
D_2 profit is maximized when the control variable is set to C_2. This offers a
better profit than that which would be obtained if the control variable was
left at C_1 as the contours are higher at Z than at Y.

The long-term steady-state optimization problem will consist of a series of
such alterations to the control variables. We can consider the problem of
maximizing the profit function over the interval $(0, T)$ as a series of maximization problems at each of a number of time points $(0, t_1, t_2, t_3, ..., T)$. For
each situation we will assume that the steady state has been reached and it is
a matter of determining the appropriate adjustment in the control variables.
Theoretically the optimum control settings should be recalculated each time
a disturbance occurs. In fact, we will select a suitable interval for reassessing
the optimum such as half an hour or two hours. The effect of the discrete
time intervals will mean that the track followed will not be as good as would

be obtained if we optimized continuously, but provided the interval is small enough it will probably be as good as is required. The discrepancy between continuous and discrete optimization is illustrated in Fig. 2.3.

We will now present three case studies of steady state optimization.

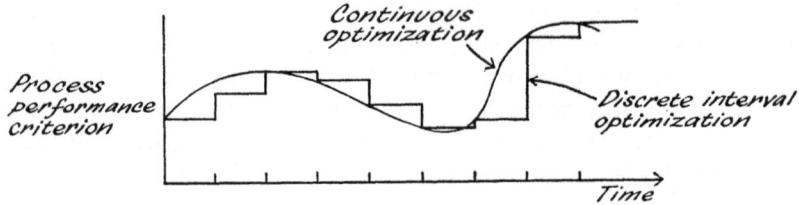

Fig. 2.3

2.3 Steady-state optimization of an oxosynthesis process

The first study of steady-state optimization reports an oxosynthesis process for the production of alcohol. Olefine, hydrogen and carbon monoxide react together in the presence of a catalyst to produce alcohol. Small quantities of paraffin and other products denoted as heavy ends are produced and some of the olefine comes through unconverted. Two impurities enter the system: paraffin in the oil feed and inert gases in the gas feed. The inputs and outputs of the process are shown in Fig. 2.4. The process is operated on a two pass system comprising a fresh feed pass and a recycle pass.

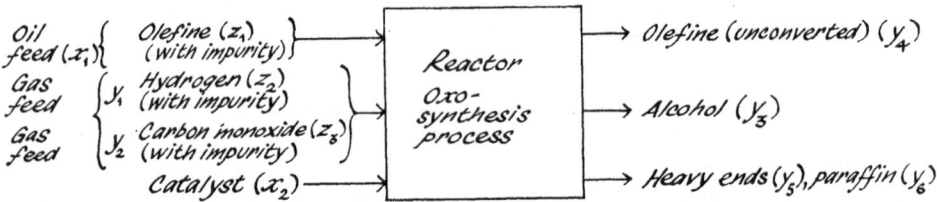

Fig. 2.4

The controllable independent variables are the oil feed rate x_1, and the catalyst feed rate x_2. The gas feed rates y_1 and y_2 are dependent variables, their values being regulated by a pressure controller to supply sufficient gas for correct operation. The other dependent variables are the output flow rates of the alcohol y_3, the olefine y_4, heavy ends y_5 and paraffin y_6. The disturbance variables are the impurities. We denote the fraction of oil that is olefine by z_1, the fraction of gas that is hydrogen by z_2, and the fraction of gas that is carbon monoxide by z_3. All the quantities x_i, y_i and z_i are functions of time.

The objective is to maximize long-term profits from the process. The profits are measured in terms of the production rates and selling prices of the products and the usage rates and cost prices of the raw materials. The production rates need to be distinguished from the feed rates as some of the materials are recycled. We use the following notation:

P_1, P_2, P_3 = production rates of alcohol, heavy ends, and paraffin,

b_1, b_2, b_3 = selling prices of alcohol, heavy ends and paraffin,

R_1, R_2, R_3, R_4 = usage rates of olefine, hydrogen, carbon monoxide and catalyst,

a_1, a_2, a_3, a_4 = purchase prices of olefine, hydrogen, carbon monoxide and catalyst,

C_1 = the service charge per unit of production of alcohol,

C_2 = the standing charge per unit time for plant operation.

Then the total profit over a period $(0, T)$ is

$$\int_0^T \left(\sum_{i=1}^3 P_i b_i - \sum_{i=1}^4 R_i a_i - P_1 C_1 - C_2 \right) dt.$$

For optimization purposes this must be expressed in terms of the controlled variables. The quantities P_i and R_i are dependent on the x_i and y_i values over time but we can use the process kinetic and mass balance equations to express the y_i in terms of x_i and thus express P_i and R_i as functions of the variables x_1 and x_2 and time t. In this form we will denote P_i by $P_i(x_1, x_2, t)$ and R_i by $R_i(x_1, x_2, t)$. Thus the total profit becomes:

$$\int_0^T F(x_1, x_2, t) dt.$$

This integral is a maximum when the expression under the integral sign is a maximum for every value of t in the interval $(0, T)$, taking account of the variation in the technical disturbance variables z_1, z_2 and z_3 and the economic disturbance variables a_i and b_i. However as it is assumed that these disturbances occur relatively infrequently the profit function is maximized by determining the values of x_1, x_2 which maximize the functions

$F(x_1, x_2, t)$

for $t = 0$, t_1, t_2, ..., T where the t_i values are appropriately chosen in the interval $(0, T)$. We call $F(x_1, x_2, t)$ for a given value of t the current objective function.

The values of the controlled variables which maximize the current objective function must satisfy some constraints. There may be constraints on the long term requirement of alcohol so that currently its production rate must lie between values p_1 and p_1'. Also the long term availabilities of olefine, hydrogen and carbon monoxide may limit their usage rates to lie between values r_i and r_i'. The plant operation may be restricted by some current constraints. Let (l_i, l_i') and (l_2, l_2') denote the bounds on the oil feed rate x_1 and the catalyst feed rate x_2 due to physical limitations. Similarly, on the output side the ratio of paraffin to alcohol y_6/y_3 may be committed to lie between L and L' imposed by distillation column restrictions. The ratio y_6/y_3 may be expressed in terms of the control variables by using process equations as

$$g(x_1, x_2, t) = y_6/y_3.$$

This completes the definition of the constraints.

The optimization problem can be solved by the created response surface technique. The form of the response surface for a given value of t will be:

$$P(x_1, x_2, r) = F(x_1, x_2, t)$$

$$+ r\left(\frac{1}{p_1' - P_1(x_1, x_2, t)} + \frac{1}{P_1(x_1, x_2, t) - p_1}\right)$$

$$+ r \sum_{i=1}^{3}\left(\frac{1}{r_i' - R_i(x_1, x_2, t)} + \frac{1}{R_i(x_1, x_2, t) - r_i}\right)$$

$$+ r \sum_{i=1}^{2}\left(\frac{1}{l_i' - x_i} + \frac{1}{x_i - l_i}\right)$$

$$+ r\left(\frac{1}{L' - g(x_1, x_2, t)} + \frac{1}{g(x_1, x_2, t) - L}\right).$$

The problem would be solved for each t-value in turn.

*2.4 Steady-state optimization of a catalytic cracking unit

In this next study we will summarize a study of an oil refinery cracking process reported by Savas. Catalytic cracking of crude-oil is an important operation in any large petroleum refinery. It is a typical process control problem. There are a relatively large number of process variables which interact in a complex manner and there are numerous constraints which must be observed. The fluctuation of the properties in the oil feed even after distillation and mixing in storage tanks cause disturbances which make the optimum operating point vary appreciably.

Fluid catalytic cracking is a continuous process for converting high-boiling, high-molecular-weight components of distilled crude oil into lower boiling, lower-molecular-weight materials such as petrol. This processing adds economic value to the oil. A powder catalyst is used to effect the reaction. The process is illustrated in Fig. 2.5. Three different streams of recycled oil x_1, x_2, x_3 are fed into the reactor together with two streams x_4 and x_5 from the storage tanks. The mass flow rates of these five feeds are x_1 to x_5. The incoming oil is vapourized on contact with the hot fluidized catalyst coming from the regeneration chamber at flow rate F_c and the cracking reaction takes place. The product vapours pass overhead and the catalyst returns to the regenerator. However the reaction deposits coke on the catalyst at rate

y_6 per unit time and this has to be burned off at rate y_7 by a stream of hot air blowing through the regenerator. The catalyst then returns to the reactor, W_c being the weight of catalyst in the reactor. The mass flow rate of air is x_6. Reactor and regenerator temperatures are given by y_1 and y_2 which are also the temperatures of the effluent gas streams. The temperatures of the five feeds are T_1 to T_5 and T_6 is the temperature of the incoming hot air stream.

The full set of variables which would need to be considered for this process is shown in Table 2.1. This may seem to be a formidable list, but it helps to

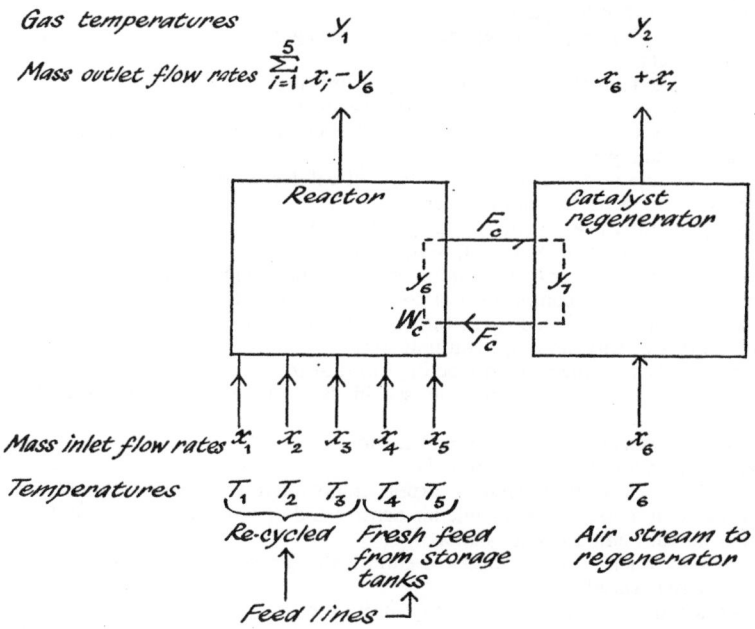

Fig. 2.5

show what a full size problem would look like. The variables x_1 to x_6 are the control variables—the inlet feed rates, y_1 to y_7 are the seven dependent variables. The disturbance variables are T_1 to T_6, H, K_0, K_1, K_2, K_3, u, n_i, A which, for any single steady state optimization are regarded as constants. The remaining quantities are known chemical and plant constants.

The objective is to maximize the production of petrol. If the fresh feed rate is constant, i.e. ($x_4 + x_5$ is constant) maximizing production is equivalent to maximizing product yield y_5 (the yield alone, without the fresh-feed constraint, is unsatisfactory as an optimization criterion because very high yields may be obtained by extensive recycling of the products at a low rate of fresh feed supply).

There are seven dependent variables $y_1, y_2, ..., y_7$ which need to be related to the control variables and the constants. We therefore need seven process equations for the model. To illustrate the form which these equations take in this case, the appropriate seven equations are listed below using the variables given in the table. Derivatives are with respect to the time variable t.

Table 2.1

Table of variables and constants

x_1, x_2, x_3	Flow rates of incoming recycle streams, weight/unit time
x_4, x_5	Flow rates of incoming fresh-feed streams, weight/unit time
x_6	Flow rate of air to regenerator, weight/unit time
y_1	Reactor temperature, °F
y_2	Regenerator temperature, °F
y_3	Light gas yield, weight/unit time
y_4	Volume-fraction oxygen in regenerator effluent gas
y_5	Product yield
y_6	Rate of coke formation in reactor, weight/unit time
y_7	Rate of burning of coke in regenerator, weight/unit time
F_c	Flow rate of catalyst between reactor and regenerator, weight/unit time
W_c	Weight of catalyst in reactor
R	F_c/W_c; inverse of average residence time of catalyst in reactor
T_i	Temperature of ith incoming stream, °F
q_i	Average specific heat of ith incoming stream, Btu/(weight)(°F)
p_i	Density of ith incoming stream, weight/unit volume
h_j	Average specific heat of jth outgoing stream, Btu/(weight)(°F)
Q_1	Exothermic heat of combustion of coke, Btu/weight of coke
Q_2	Endothermic heat of reaction, Btu/(weight)(°F)
Q_3, Q_4	Heat losses to surroundings, Btu/unit time
Q_5	Net heat from regenerator to reactor, Btu/unit time
m_1, m_2	Constants representing amount and specific heat of material in reactor and regenerator, respectively
M_h, M_c	Molecular weights of hydrogen and carbon
H	Weight fraction of hydrogen in coke
k	Mole ratio of carbon dioxide to carbon monoxide in regenerator effluent gas
a, b, c	Various constants, which include air density and composition factors and miscellaneous empirical constants
K_0, K_1, K_2 $\left.\right\}$ K_3, u, n_i, A	Constants which must be evaluated empirically for each different oil stock (disturbance variables)
C	Coke accumulation in system, weight/unit time

The first equation is the reactor heat balance:

$$m_1 \frac{dy_1}{dt} = \sum_{i=1}^{5} q_i T_i x_i - \left(\sum_{i=1}^{5} x_i - y_6 \right) y_1 h_1 - Q_2(y_1 + a) \sum_{i=1}^{5} x_i + Q_5 - Q_3.$$

The second equation is the overall heat balance equation:

$$\frac{dQ_5}{dt} = m_1 \frac{dy_1}{dt} + m_2 \frac{dy_2}{dt} = \sum_{i=1}^{6} q_i T_i x_i - \left[\left(\sum_{i=1}^{5} x_i - y_6 \right) y_1 h_1 + (x_6 + y_7) y_2 h_2 \right]$$
$$+ Q_1 y_7 - Q_2(y_1 + a) \sum_{i=1}^{5} x_i - (Q_3 + Q_4).$$

The overall coke balance equation is

$$\frac{dC}{dt} = y_6 - y_7.$$

These first three equations have been expressed as dynamic equations, but in the steady state the left-hand sides will be set to zero. The steady-state coke-burning defines the coke burning rate in terms of the regenerator air rate, the oxygen level in the flue gas, the coke composition and various fixed parameters. This provides the equation:

$$y_7 = \frac{(b - cy_4)x_6}{(cy_4 + 1)\dfrac{H}{M_h} + \dfrac{(1-H)}{M_c}\left(\dfrac{2k + cy_4 + 1}{k + 1}\right)}.$$

The reaction rate equation for the formation of coke provides a fifth equation as

$$y_6 = \frac{K_0 F_e e^{-A/y_1}}{\displaystyle\sum_{i=1}^{5} x_i} \sum_{i=1}^{5} x_i \frac{g(h_i + 1)}{R^{n_i + 1}}$$

where $g(a)$ is the gamma function $\displaystyle\int_0^\infty x^{a-1} e^{-x} dx$.

The product yield equation

$$y_5 = K_1 \left(\frac{y_6}{\displaystyle\sum_{i=1}^{5} x_i}\right)^u \sum_{i=1}^{5} \frac{x_i}{p_i}.$$

The by-product yield equation provides the final relationship:

$$y_3 = K_2 y_5 (y_1 + K_3).$$

Finally we need to note the control and quality constraints on the x_i and y_i values. The flue-gas oxygen level y_4 is of critical importance and is subject to both upper and lower limits. There are upper limits on the rate of coke formation. These constraints on the dependent variables may be written:

$$a_i \leq y_i \leq b_i.$$

There will also be the usual constraints on the control variables caused by limitations on the size of pipes, tanks, compressors, etc. These will take the form:

$$A_i \leq x_i \leq B_i.$$

We now have a set of problem variables x_i and y_i, an objective function and a set of equality and inequality constraints expressing the full optimization problem. It is one of the most complex problems we have met so far. The best method for solving the problem would be the created response surface or the direct search procedure.

2.5 The optimization of lime process manufacture

The third study in steady-state optimization will describe how the manufacturing process in a lime factory can be optimized. The process essentially consists of a series of screens and classifiers for separating off the different grades of lime which can be obtained from the initial material.

The components of the process are illustrated in the Fig. 2.6. The lime is burnt in a lime kiln and screened in a grate under the kiln. The lump lime that does not come through the grate will be called product 1. The high-grade kiln fines which go through the grate are separated off and provide product 2. The low-grade fines which go through the grate are further screened —the low-grade unslaked fines coming through the screen forming product 3. The material passing over the screen is put through a hydrator and subsequently through classifiers in series giving two fine hydrate products 4 and 5. The remainder from the classifiers is milled giving a fine low-grade product 6.

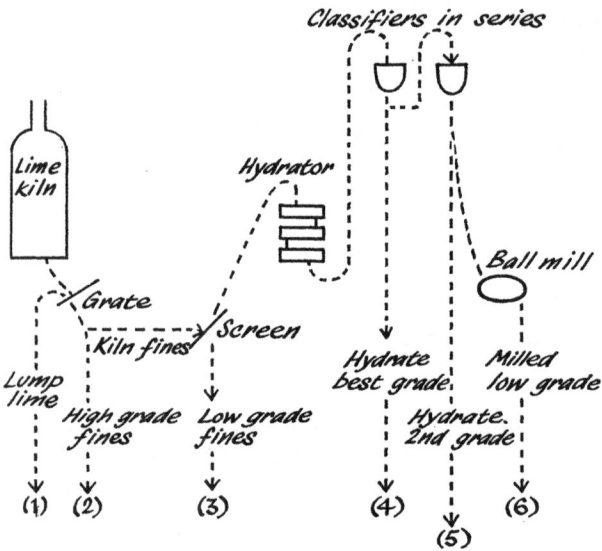

Fig. 2.6

It is assumed that at any given time we know the quantities of the different products which will be produced at each stage from the initial raw material with the kiln working at normal capacity.

The products need not all go into saleable items or be disposed of by dumping. Some of the products can be reprocessed and exchanged into different products. There are six possible reprocessing transfers of this kind. The high-grade fines, product 2, may be conveyed to a ball mill and sold as product 6. The saleable lump lime product 1 can be diverted to low-grade fines, thus giving additional products 3, 4, 5 and 6. Lump lime 1 can alternatively be broken up and sold as high-grade fines product 2. The first classifier can be by-passed so that product 4 becomes products 5 and 6. Or by-passing the second classifier means product 5 is converted into product 6. Finally, product 3 may be passed through a classifier to give product 6. We can express these six possible reprocessing transfers by means of a matrix. Denote the six transfers by T_1, T_2, ..., T_6, and let t_{ij} denote the quantity of product j obtained from a unit input in transfer i. Then the matrix given in Table 2.2 gives all the information for the transfers which have been described. A minus-one entry in the matrix indicates a unit of input of a product. For example, in T_1 a unit input of product 1 produces t_{16} units of product 6. This completes the description of the process.

Table 2.2

Transfers

m	T_1	T_2	T_3	T_4	T_5	T_6
Product 1		-1	-1			
2	-1		t_{32}			
3		t_{23}				-1
4		t_{24}		-1		
5		t_{25}		t_{45}	-1	
6	t_{16}	t_{26}		t_{46}	t_{56}	t_{66}

The control variables are the quantities which should be put through each exchange process, the amount which is sold and the amount dumped. We will define the following notation and data for any given period of operation:

x_j = quantity to be put through transfer process T_j

s_i = quantity of product i made for sales

d_i = quantity of product i made to be dumped. (Dumping is necessary if the product cannot be sold)

n = number of kilns

b_i = quantity of product i which is produced by 1 kiln

r_i = orders for product i

p_i = revenue from sale of unit quantity of product i

h = cost of operating with 1 kiln

k_j = cost of putting a unit quantity through transfer process T_j

c_i = cost of dumping product i

The problem variables are x_j, s_i and d_i.

We wish to maximize the total profit. This is measured as the total return from sales less the cost of dumping and the costs of production, i.e.

$$F(X, S, D) = \sum_{i=1}^{6} (s_i p_i - d_i c_i) - \sum_{j=1}^{6} k_j x_j - hn.$$

The process equations express the requirement that the total production of a product from the kiln plus the exchange processes must equal the amount sold and dumped. There is one equation for each product using the table of exchanges:

$$b_1 n - x_2 - x_3 = s_1 + d_1$$

$$b_2 n - x_1 + t_{32} x_3 = s_2 + d_2$$

$$b_3 n + t_{23} x_2 - x_6 = s_3 + d_3$$

$$b_4 n + t_{24} x_2 - x_4 = s_4 + d_4$$

$$b_5 n + t_{25} x_2 + t_{45} x_4 - x_5 = s_5 + d_5$$

$$b_6 n + t_{16} x_1 + t_{26} x_2 + t_{46} x_4 + t_{66} x_6 = s_6 + d_6.$$

Finally, we have the inequality constraints to ensure that the variables are positive and that sales are less than orders,

$$x_j \geqq 0$$

$$s_i \geqq 0$$

$$d_i \geqq 0$$

and

$$s_i \leqq r_i.$$

This completes the specification of the optimization problem. All the functions are linear, and it can be solved by the simplex method.

It may be noted that the number of kilns n may be treated as a variable, and as this enters into the problem in a linear manner, the same optimization technique can be used if it is desired to include this factor as a problem variable.

*2.6 The dynamic optimization problem

Up to this point we have investigated the optimization of chemical processes in terms of a steady state solution: we have assumed that when disturbances occur, we should redetermine the control settings which maximize profit and simply implement these new control values. This policy is practicable if the disturbances are relatively infrequent and the process responds fairly rapidly to control action. At least we assume that the current values of the disturbance variables will persist until the process has reached the new steady state optimum. As the disturbance frequency increases, or as the process response time falls the process will spend less and less time at the optimum. In these situations where the transient behaviour is economically significant the dynamic manipulation of the control variables over time is necessary to achieve improved performance. A particularly important field for the dynamic optimization of the controls arises in batch production processes where the reaction is required to convert each batch from its initial state or composition to its required final composition. A typical objective in batch production is to achieve the conversion of the raw material in the minimum time. The reaction converting the material into the products can probably take place at a variety of temperatures and pressures. However, by varying the controls over the course of the reaction we may significantly reduce the time required for the conversion to take place. This in turn leads to a cost saving. Material and energy costs in a batch process are generally independent of processing conditions. The major controllable costs are labour, overheads and the plant investment: all of these increase with processing time so that the return on these can be improved by reducing the time required per batch. In a batch hydrogenation process reported by Eckman and Lefkowitz a reduction of 23 per cent in processing time was achieved by the use of optimization techniques.

The total processing time for the batch may be expressed as an integral of a function of the composition variables and the parameters of the reaction expressing the state of the reaction over time. As we are dealing with changes over time the representation of the process involves differential equations which have not yet been met in any previous optimization problem. The minimization of these 'time' integrals in which derivatives are involved has been examined in the mathematical field known as the calculus of variations, but we will concentrate here on the use of the method of dynamic programming for solving minimum time path problems. In the next section we will review a study by Roberts and Mahoney which fully illustrates the technique.

2.7 Dynamic programming control of a batch reaction

We will begin by describing the way in which a particular reaction is defined dynamically. A feed stock of three components X, Y and Z at concentrations (x_0, y_0, z_0) is to be converted into a final product of composition (x_f, y_f, z_f) in the minimum possible time. The conversion process takes place in a reactor: X is converted into Y at rate K_1 and Y into Z at rate K_2, and also back into X at rate K_2. Symbolically we can denote this reaction scheme by the arrow diagram of Fig. 2.7.

$$X \underset{K_3}{\overset{K_1}{\rightleftharpoons}} Y \overset{K_2}{\longrightarrow} Z$$

Fig. 2.7

As we are dealing with rates of change over time, the descriptive process equations must be expressed in terms of derivatives with respect to time.

If x, y and z denote the concentrations of X, Y and Z at any time t in the course of the reaction, the conversion process can be described by the differential equations:

$$\frac{dx}{dt} = -K_1 x + K_3 y$$

$$\frac{dy}{dt} = +K_1 x - (K_2 + K_3) y$$

$$\frac{dz}{dt} = K_2 y$$

$$x + y + z = 1.$$

The first equation describes the fact that the rate of increase of x is proportional to its decrease rate K_1 and its creation through y at rate K_3. Similar descriptions can be given for the other equations. The quantities K_i are known as the reaction velocity constants and they are functions of pressure and temperature. Here we consider the reactions to occur isothermally and the K_i are functions of pressure alone. The last equation shows that the sum of concentrations is unity and if any two of the variables are determined the third quantity is known. The set of differential equations therefore describes the way in which x, y, and z would change as time increases depending on their current values and the pressure settings as expressed in the K_i quantities. It should be noticed that so far we have no explicit time variable in the problem.

Suppose we start off with an initial composition x_0, y_0, z_0, and assume that there is a given initial pressure P_0. We want to alter the composition to a final specification x_f, y_f, z_f (there being no final pressure given). One policy would be to leave the pressure steadily at P_0 and let the reaction take place trusting it will reach the composition x_f, y_f, z_f at some future time. But we are allowed to control the pressure and this offers various possible ways of organizing the reaction. The possibilities are constrained so that the final composition is x_f, y_f, z_f, but within this limitation the pressure is to be manipulated so as to achieve the conversion in the minimum time. The problem variables are, in effect the pressure at each moment in time, the objective function is the time of conversion which is a function of the pressure values, and the constraints require that the initial pressure is P_0 and that the total sequence of pressures will lead up to the correct final composition. Also the composition and pressure are interrelated over time through the process differential equations.

The idea of a pressure policy can be illustrated graphically. As the three compositions are interdependent through the equation

$$x + y + z = 1$$

the composition can be specified at any time by giving values to two of the variables. Let us take the two variables x and z. Then starting from the initial point (x_0, z_0), we need to move to the point (x_f, z_f) in the minimum time. Clearly there are many possible paths each with its own reaction time.

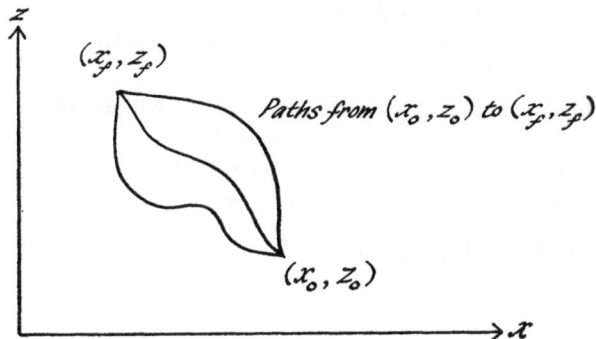

Fig. 2.8

Three possible paths are shown in Fig. 2.8. The different paths are distinguished by the various ways in which the pressure could be manipulated in the course of the reaction.

Now to organize the determination of the minimum time path in a suitable form for dynamic programming we superimpose an arbitrary slanting grid

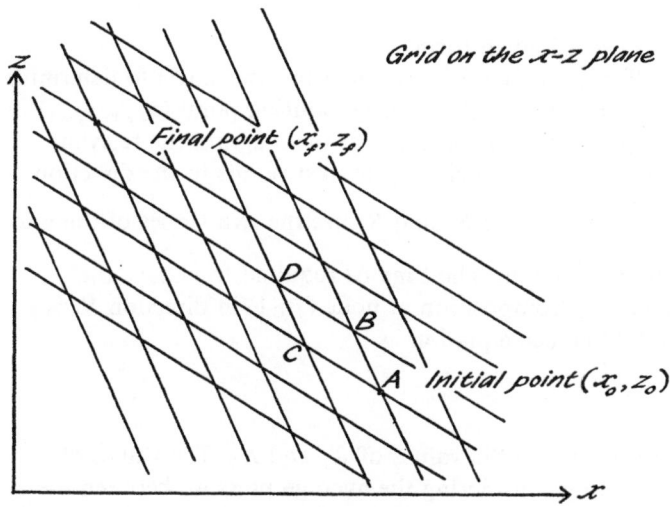

Fig. 2.9

over the x-z plane (as shown in Fig. 2.9) and restrict the number of paths which we are going to consider to the set of routes passing from the initial point to the final point through the grid points called nodes. (The grid is slanting as the derivatives $\dfrac{dx}{dz}$ cannot be zero since this would mean that a steady state equilibrium had been reached.) In moving from the initial to the final point we can see that at each node we have the choice of moving in one of two directions, either slanting upwards or along. If we can measure the time taken to move along any leg of the grid between two adjacent nodes and assess the pressure changes required to implement the movement, then we are faced with a shortest time path problem which is readily tackled by dynamic programming.

The area which the grid must cover is clearly limited and we will assume that there are M nodes with coordinates (x_j, z_j) for $j = 0, 1, ..., M$. The grid system imposes a discrete nature on the problem and we therefore convert the original differential equations into difference equations and write the equations for $\dfrac{dx}{dt}$ and $\dfrac{dz}{dt}$ at the point (x_j, y_j, z_j) as

$$\frac{x_{j+1} - x_j}{\Delta t} = -K_1 x_j + K_3 y_j$$

$$\frac{z_{j+1} - z_j}{\Delta t} = K_2 y_j$$

where Δt is a small time increment. (Note that the equations can be expressed entirely in terms of x_j and z_j as $y_j = 1 - x_j - z_j$.) Also let us define the functions of pressure K_1, K_2, K_3 as

$$K_i = h_i(P), \; i = 1, 2, 3$$

which are assumed to have a known form.

We will use these difference equations and the functions $h_i(P)$ to determine the time to move from a point (x_j, z_j) to a neighbouring point (x_{j+1}, z_{j+1}) on the grid. If the index j referred to the point A in the diagram the relevant neighbouring points are B and C which are the next points in the direction of slope $\dfrac{dz}{dx} = S_1$ or $\dfrac{dz}{dx} = S_2$, where S_1 and S_2 are the two slopes of the grid in the directions to the next points. The time Δt required to change the composition at node j to the composition at node $(j+1)$ in direction S_1 is obtained from the first difference equation as

$$\Delta t = \frac{x_{j+1} - x_j}{-K_1 x_j + K_3 y_j}.$$

Δt will be determined if we know the values of K_1 and K_3. The values of K_1 and K_3 are determined by considering the average pressure between the two nodes j and $(j+1)$. As we are proceeding in the direction S_1, the following equation can be obtained by dividing the two difference equations:

$$S_1 = \frac{z_{j+1} - z_j}{x_{j+1} - x_j} = \frac{K_2 y_j}{-K_1 x_j + K_3 y_j}.$$

We want to choose values of the pressures at nodes j and $(j+1)$ to satisfy this equation.

If we work outwards systematically from the initial point where the pressure P_0 is given, we will calculate the pressure P_j at node j before reaching node $(j+1)$ with unknown pressure P_{j+1}. If we force the last equation to hold at the average pressure $\frac{1}{2}(P_j+P_{j+1})$ between nodes j and $(j+1)$ it can then be written as

$$\frac{h_2[\frac{1}{2}(P_j+P_{j+1})]y_j}{-h_1[\frac{1}{2}(P_j+P_{j+1})]x_j+h_3[\frac{1}{2}(P_j+P_{j+1})]y_j} = S_1$$

and, knowing P_j, we can solve this equation for P_{j+1}. Hence the K_i values can be determined at the average pressure as $h_i[\frac{1}{2}(P_j+P_{j+1})]$ and these values can be substituted in the equation for Δt. We have thus found the time required to move from node j to node $(j+1)$, and the pressure P_{j+1} to be applied at node $(j+1)$. This time will be denoted by $\Delta T_i(P_{j+1})$ with the suffix $i = 1$ or 2 according to whether we are proceeding in the directions S_1 or S_2 to reach node $(j+1)$.

The dynamic programming procedure can now be applied to find the shortest time through the network knowing the times on the links in the grid. Let $f_k(x_j, z_j, P_j)$ be the minimum time for the system to move from the initial point (x_0, z_0) with pressure P_0 to the point (x_j, z_j) at pressure P_j in k steps. Then the following recurrence relations hold

$$f_k(x_j, z_j, P_j) = \min \begin{Bmatrix} \Delta T_1(P_j)+f_{k-1}(x_i, z_i, P_i) \\ \Delta T_2(P_j)+f_{k-1}(x_l, z_l, P_l) \end{Bmatrix} \text{ for } k = 0, 1, 2, ..., M.$$

It states that in moving to point (x_j, z_j) we must come from the points (x_i, z_i) or (x_l, z_l) adjacent to (x_j, z_j) and which reach (x_j, z_j) by moving in the directions S_1 and S_2. For example, in Fig. 2.9 we must reach node D from nodes B or C. The direction of movement and the pressure P_j in $f_k(x_j, z_j, P_j)$ are noted for whichever route is better. The final composition (x_f, z_f) is reached in time $f_M(x_M, z_M)$ and we can determine the route which has led to the minimum time path by tracing back through the nodes and the pressures as in a shortest route calculation. We thus determine the pressure control plan to adopt so as to minimize the batch conversion time. It should be noted that in the dynamic programming recurrence relations we can readily ensure that any bounds on the pressure value are satisfied by excluding any point which requires a pressure P_j violating these bounds.

Example 2.4

Roberts and Mahoney have worked out an example using this dynamic programming scheme and the reaction system described above. Suppose the initial and final points are

Initial point	Final point
$x_0 = 0.88000$	$x_f = 0.6000$
$y_0 = 0.11365$	$y_f = 0.29765$
$z_0 = 0.00635$	$z_f = 0.10235$

The initial pressure is 298 and a grid with slopes $S_1 = -0.3$ and $S_2 = -0.4$ was employed. The constants K_i were expressed as functions of pressure:

$$K_1 = 1.0P^{-5}$$

$$K_2 = 5(10^{-4})P^2$$

$$K_3 = 1.2594.$$

The 25 grid points were as shown in Fig. 2.10 marked with letters a to y

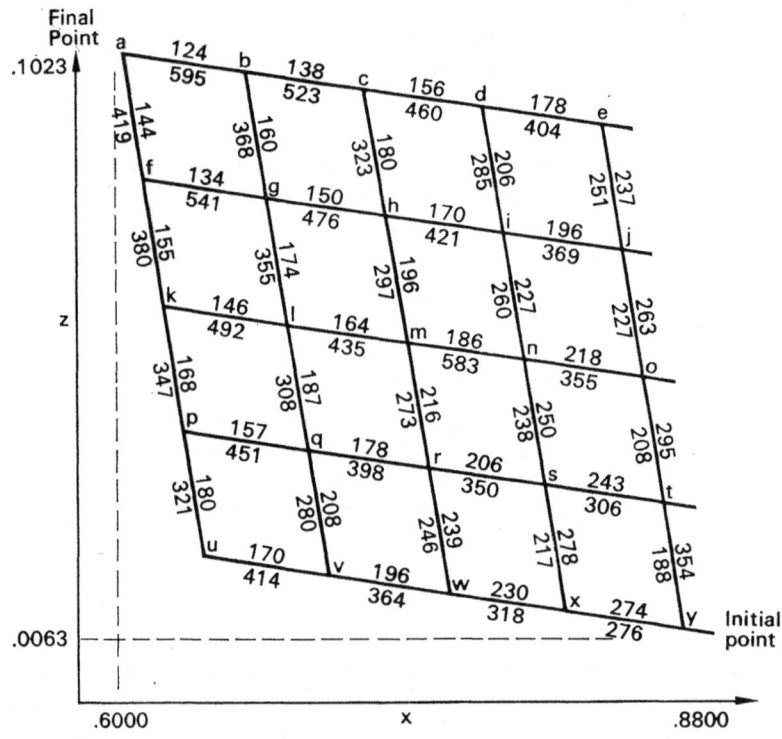

Fig. 2.10

and numbers marked on the network give the times of travel between adjacent nodes and the average pressure to drive the composition between the nodes. The numbers below the lines or to the left represent the time of travel, and the numbers above the lines or to the right represent the average pressure between the adjacent nodes.

There are two optimal policies expressed in the two sequences of nodes:

$v, x, w, v, q, p, k, f, a$

$v, x, w, r, q, p, k, f, a.$

2.8 The optimization of multi-stage processes

So far we have considered the steady-state problem and the task of producing a batch of the required product in the minimum time. A further important field for the application of optimization techniques in chemical process control occurs in the optimization of multi-stage systems such as arise in distillation columns and series of stirred tank reactors. Here the problem is to determine the control settings at a number of stages where the output from the nth stage forms the input to the $(n+1)$th stage. Dynamic programming is once again the technique which is used for solving these problems. The procedure for maximizing profits over all stages usually fits the multi-stage scheme

$$
\begin{bmatrix} \text{Maximum} \\ \text{profit} \\ \text{from} \\ N \text{ stages} \end{bmatrix} = \text{Maximum} \left[\begin{pmatrix} \text{Profit} \\ \text{from} \\ \text{first} \\ \text{stage} \end{pmatrix} + \begin{pmatrix} \text{Maximum profit} \\ \text{from remaining} \\ (N-1) \text{ stages} \\ \text{with feed} \\ \text{produced by} \\ \text{first stage} \end{pmatrix} \right].
$$

It may be written alternatively in terms of the minimization of cost.

We will illustrate the optimization of such multi-stage processes by considering a batch-reflux process examined by Mitten and Prabhakar. In the exercises a holding-time problem over a series of reactors is also investigated.

Example 2.5 Optimization of a batch-reflux process
A batch of feed stock of a given quantity and composition is to be distilled in a given piece of equipment consisting of a reboiler, a column, a condenser and reflux apparatus. The material is passed through the series of equipments some of it being distilled on each pass. Throughout the distillation, the reflux ratio may be periodically adjusted within defined physical limits, subject to the requirement that the total distillate collected must be of the given composition and amount. The set of reflux ratio adjustments which gives the most economical operation of the equipment must be found.

Suppose there are b_0 moles of feed stock which are required to produce a batch of d moles of distillate, and that the composition of the L constituents of the feed stock is $X_0 = (x_{01}, ..., x_{0L})$ and of the required distillate is $Y = (\bar{y}_1, ..., \bar{y}_L)$ where x_{0i} and $\bar{y}_i$ are the mole fractions of component i in the feed and in the total distillate. The batch distillation is to be carried out on a series of N stages, numbered 1, 2, ..., N during each of which a specified number of moles of distillate is to be collected. The quantity to be collected at stage n is d_n (which is given), and we therefore require that $\sum_{n=1}^{N} d_n = d$.

The reflux ratio r_n at stage n can be altered from stage to stage, but it is held fixed throughout a given stage. The cost of operating the plant depends on the reflux ratios chosen, these quantities representing variable operating costs, reboiler heat requirements and other factors.

Let b_n be the total moles that remain at stage n in the reboiler and let their composition be $X_n = (x_{n1}, x_{n2}, ..., x_{nL})$. Let $Y_n = (y_{n1}, y_{n2}, ..., y_{nL})$ be the composition of the d_n moles of distillate collected during stage n. Then the total and component materials balance over each stage may be written as

$$b_n = b_{n-1} - d_n$$

and

$$b_n X_n = b_{n-1} X_{n-1} - d_n Y_n$$

for $n = 1, 2, ..., N$. It is assumed that knowledge of the distillation process and the equipment enable the nth stage composition Y_n to be computed given the composition of the reboiler liquid X_n, the reflux ratio r_n and the amount of distillate removed d_n. We therefore have a functional relation of the form:

$$Y_n = H(b_n, X_n, d_n, r_n).$$

Using the previous two equations we can express a relation between X_n and X_{n-1} as

$$b_{n-1} X_{n-1} = b_n X_n + d_n H(b_n, X_n, d_n, r_n).$$

As the d_n quantities are given, and therefore by the total material balance the b_n quantities are known, X_{n-1} can be expressed simply in terms of the unknowns r_n, X_n as

$$X_{n-1} = h_n(r_n, X_n).$$

Thus we can compute the reboiler composition at the beginning of each stage, if one is given the composition at the end of the stage and the reflux ratio used during the stage.

The cost of operating the equipment is assumed to depend only on the reboiler conditions at the beginning and end of the stage, on the reflux ratio used during the stage, and on the amount and composition of distillate collected during the stage. Thus is C_n denotes the cost during stage n, C_n may be expressed in function form as if

$$C_n = G(b_{n-1}, X_{n-1}, b_n, X_n, r_n, d_n, Y_n)$$

which, using the above relations may be written in the variables as

$$C_n = g_n(r_n, X_n).$$

The operating costs over the entire N-stage process is

$$C(r_1, r_2, ..., r_N) = K + \sum_{n=1}^{N} g_n(r_n, X_n)$$

(where K is an overheads constant), and we wish to minimize this function taking account of the successive constraints on the X_n values.

For the dynamic programming solution, let us define $f_n(X_n)$ as the minimum cost obtainable for an n-stage process with output X_n. The various values of

X_n define the states of the system at stage n, and the means of reaching state X_n from X_{n-1} depend on the reflux ratios chosen. Thus the recurrence relations may be written as

$$f_n(X_n) = [\min_{r_n} (g_n(r_n, X_n) + f_{n-1}(X_{n-1}))].$$

For any given value of X_{n-1} we are committed to an appropriate reflux ratio r_n by the relation $X_{n-1} = h_n(r_n, X_n)$. The recurrence relation can be rewritten to incorporate this constraint as:

$$f_n(X_n) = \min_{r_n} [g_n(r_n, X_n) + f_{n-1}(h_n(r_n, X_n))].$$

This is a single minimization in terms of the variable r_n, and it can be evaluated for an appropriate set of values of X_n in the order $n = 1, 2, ..., N$ where $f_0(X_0) = 0$. At each stage, and for each state, we would record the optimal reflux ratio to adopt, say $r_n^*(X_n)$.

Mitten and Prabhakar have applied this procedure to the distillation of benzene from a mixture of benzene and toluene in a batch fractionating column. Their process has four stages and the objective function is simply the sum of the reflux ratios. The optimal results are compared with two policies commonly employed in practice in which either the reflux ratio is continuously varied to maintain a uniform product composition throughout, or the reflux ratio is held constant. In the former case the optimum policy achieved by dynamic programming offers a 50 per cent saving and in the latter a 5 per cent saving in operating cost.

REFERENCES

Aris, R. 1964. *Discrete Dynamic Programming*. Blaisdell.
Aris, R. 1964. *The Optimal Design of Chemical Reactors*. Academic Press, New York.
Chamoux, J. P. 1967. Optimization de la synthese sulphurique. *Rev. fr. Inform. Rech. op.*, **1** (6).
Chien, K. L. 1961. Computer Control in Process Industries. Chap. 20 in *Computer Control Systems Technology*. Ed. C. T. Leondes. McGraw-Hill, New York.
Eckman, D. P. and Lefkowitz, I. 1960. *Principles of Model Techniques in Optimizing Control. Proc. First Congress Int. Fed. Autom. Control Moscow*. Butterworth, London.
Eckman, D. P. and Lefkowitz, I. 1957. Optimizing control of a chemical process. *Control Eng.*, **4** (9), 197.
Fiore, C. F. and Rozwadowski, R. T. 1968. The implementation of process models. *Man. Sci.*, **14** (6).
Gould, L. and Kipiniak, W. 1960. Dynamic optimization and control of a stirred tank chemical reactor. *Trans. A.I.E.E.* I, **79**, 734.
Himmelbau, D. M. 1963. Process optimization by search techniques. *Ind. Eng. Chem. Process Design Development*, **2**, 296.
Lefkowitz, I. and Eckman, D. P. 1959. Optimizing control by model methods. *Instrument Soc. Am. J.*, **6** (7), 74.
Lupfer, D. E. and Johnson, M. L. 1963. Automatic control of distillation columns to achieve optimum operation. *Proc. Joint Autom. Control Conf.*, 145.
Mitten, L. G. and Prabhakar, T. 1955. Optimization of batch-reflux processes by dynamic programming. *Chem. Eng. Prog. Symp.*, Series No. 50, **60**, Am. Inst. Chem. Eng.
Roberts, S. M. and Lyons, H. I. 1961. The gradient method in process control. *Ind. Eng. Chem.*, **53**, 877.

Roberts, S. M. and Mahoney, J. D. 1962. Dynamic programming control of batch reaction. *Chem. Eng. Progr. Symp. Series*, **58** (37), 1-9.
Robertson, H. H. and O'Grady, W. P. 1966. Steady state optimization of an oxosynthesis process. *Proc. Int. Fed. Autom. Control*, London.
Savas, E. S. 1965. *Computer Control of industrial Processes*. McGraw Hill, New York.
Shrage, R. W. 1959. Optimizing a catalytic cracking operation by the method of steepest ascents. *Opr. Res.*, **6** (8), 94.
Wilde, D. J. 1965. Strategies for optimizing macrosystems. *Chem. Eng. Progr.*, **61** (3), 86.

Exercises on chapter 2

1 A chemical reactor has two control variables x_1 and x_2. x_1 is the flow rate of a liquid reactant and x_2 is the pressure of a gaseous reactant. The maximum flow rate of x_1 is 60 units per unit time because of pipe and compresser capacities and the maximum permitted level of x_2 is 40 units for safety reasons. The profit P obtained from the product is a linear function of the two control variables

$$P = -50 + 6x_1 + 5x_2.$$

A by-product impurity is formed in the reaction and it cannot be tolerated beyond a level of 108 units. The expression for the by-product output in terms of the control variables at one time was

$$x_1 + 2x_2$$

but owing to a change in the purity of the feed it has now become

$$\tfrac{5}{3}x_1 + x_2.$$

Determine the alteration in the optimal operating control settings due to this disturbance and measure the change in the profit level.

2 A chemical process has two feed lines which yield a marketable product and a by-product. If the flow rates of the two feed lines into the reactor are z_1 and z_2 the output rate of the product is $h(z_1, z_2)$. (A mass balance will determine the rate of output of the by-product.) The product and by-product can either be sold at known prices or the by-product may re-enter the process on the second feed line after regeneration. The feed stocks have known costs and the cost of regeneration is also known. The flow rates of the initial feed lines are constrained to lie between certain limits.

   Define the variables for this process and formulate the problem of determining the inlet rates so as to maximize profits.

3 A refinery uses four primary components—alkalytes (A), catalytically cracked gasoline (C), straight-run gasoline (S) and isopentane (I)—to produce three marketable products: aviation fuel (N) of minimum performance number 91, aviation fuel (H) of minimum performance number 100 and premium petrol (P). The price for one unit of N is £5, for one unit of H is £6 and for one unit of P is £4. The primary components are in limited supply (and costs may be neglected), the supply of A, C, S and I being 38, 25, 40 and

14 per unit time. The performance number of a mixture of A, C, S and I is obtained by giving weights 107, 93, 87 and 108 respectively to the proportions of the components. Also it is required that the vapour pressure of the mixtures of N and H be kept below 7 lb in^2, the pressure being measured by attaching weights 5, 8, 4 and 20 to the proportions of the components A, C, S and I. Formulate the problem of determining how much of each component should go into each product per unit time to maximize revenue assuming that all the material is used.

4 A manufacturing process produces a range of four products which are various grades of a powder. The initial raw material passes through a classifier to produce product 1, the remainder goes through a further classifier to produce product 2 and the rest goes through a third classifier to separate products 3 and 4. The quantities of each product produced from 1 ton of raw material are 4, 6, 7, 3 cwt respectively. However, product 1 can be further processed by a grinder and re-entered at classifier 2 to produce proportions 1 : 2 : 1 of products 2, 3 and 4 independently of the main stream. Product 2 can also be processed on the grinder and entered at classifier 3 to produce a ratio of 2 : 3 units of products 3 and 4.

The orders for products 1, 2, 3, 4 are 10, 24, 15, 18 tons respectively. 50 tons of raw material are available to meet these orders. The revenue from the products are, in £/ton, 7, 10, 16, 25. The cost of putting 1 ton of products 1 and 2 through a grinder are £5, £10 respectively. The cost of dumping 1 ton of any product is £5.

Formulate the problem of determining the amount of each product to sell and to be dumped, and the amounts of products 1 and 2 to be re-processed so as maximize the total profit. It is assumed that the process has adequate capacity.

5 A batch reaction mixture is made up of three chemical components identified as X, Y and Z. A gas under pressure reacts with X and Y so that X transforms into Y at rate K_1 and Y transforms into Z at rate K_2 as shown in the scheme:

$$X \xrightarrow[K_1]{} Y$$
$$Y \xrightarrow[K_2]{} Z.$$

The kinetic reaction coefficients K_1 and K_2 are known functions of pressure. K_1 is directly proportional and K_2 inversely proportional to the pressure, following the lines considered in Section 2.7. Write down the differential and difference equations for this reaction and present the formal dynamic programming relationship for manipulating pressure to minimize the batch reaction time. It is assumed that the initial and final compositions are known as well as the initial pressure, but the pressure throughout is constrained to lie between the values L_1 and L_2.

6 The energy required to compress a gas from pressure p_0 to pressure p in N stages is proportional to the expression

$$E = \left(\frac{p_1}{p_0}\right)^q + \left(\frac{p_2}{p_1}\right)^q + \ldots + \left(\frac{p_N}{p_{N-1}}\right)^q$$

where p_i is the pressure at the ith stage, and q is a constant. Show how to choose the intermediate pressures $p_1, \ldots, p_N$ so as to minimize the energy requirement, and calculate their values for a three stage system.

*7 A process stream enters a series of stirred tanks numbered 1 to N. The feed stock A is to be converted to B by allowing the reaction to take place over the N stages. The concentration of B in the inlet stream is C_0 and the required concentration in the product is C_N. The extent of the reaction in the successive stages depends on the holding time within each tank, and the output concentration from the previous tank. The concentration C_n at the end of stage n is given by the relation

$$C_n = \frac{C_{n-1} + h_n K_1}{1 + h_n(K_1 + K_2)}$$

where h_n is the holding time in tank n and K_1, K_2 are reaction coefficients. Show how to determine the series of holding times $h_1, h_2, \ldots, h_n$ so as to minimize the total holding time. Also show that for a two-stage system the optimal policy requires the holding times to be equal.

*8 It is desired to separate four different particle sizes of coal by means of industrial screening equipment. The equipment operates in the following manner. The raw feed which contains all four sizes is placed on a vibrating screen. The largest particles (size 3) are retained on the screen, but the small particles (sizes 0, 1, 2) pass on to the next screen. However, the amount of smaller particles passing through the screen depends upon the degree and time

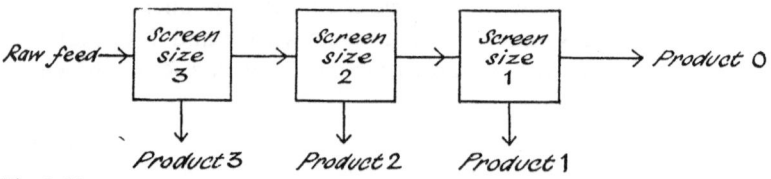

Fig. 2.11

of vibration. The process continues in this manner, passing through 3 screens, to produce 4 products. The material retained on each screen is removed for sale. The last product which is labelled product 0 is obtained from the material passing through the third screen. A diagram of the process is given in Fig. 2.11. The amount of the undersize material that is passed on to the next screen is given by the formula:

(amount passed to next screen i) = z_i (amount that entered the screen i)

where $0 \le z_i \le 1$, and depends upon the degree and time of vibration.

The cost of separation at each screen depends on the z_i values and on the total amount of material entering the screen. This relationship is given as follows:

(separation cost for the ith screen)

$$= \text{(amount of material entering the } i\text{th screen)} \cdot k_i \log (1 - z_i)$$

where k_i is a negative constant.

The feed contains l_i tons of product i material. The product removed from screen i sells at $£p_i$ per ton. Show how to determine the composition of the product removed for sale at each stage and to find the z_i values to maximize profit.

*9 An important problem in the coordinated running of electric power plants is to allocate a given power load among a number of generating units so as to minimize the total cost of supplying power. If there are N generating stations and the cost of supplying power x_i from station i is $f_i(x_i)$, and the transmission loss is given by $L(x_1, x_2, ..., x_N)$ determine the conditions of optimum operation to meet a total demand D.

3 Machine sequencing and resource utilization

3.1 Short-term planning in a factory

There are a number of problems in a factory of a continuing nature which arise every week or every fortnight or even every day. Their frequency means that only a small improvement in methodology could lead to substantial annual savings. The most important of these problems is the efficient loading of machines with the current workload so that the various jobs in the factory are being progressed at the correct rate and the full resources of the factory are being properly organized. Competent short-term planning of this kind is particularly relevant in engineering jobbing shops and batch manufacturing processes where set-up costs are high, and the ratio of the value added by processing to the material value of the product is comparatively low, as it is then important to sequence the jobs efficiently, keep in-process stocks down and avoid too much stock-holding.

The production planning problem can often be reduced to the task of deciding in what order the operations or batches should be processed through the machines. The total set of operations to be scheduled are determined by the current collection of customer orders on the factory. Each order or job is broken down into a series of processing operations dictated by the technical specifications for manufacturing the product. For example, a job may have to be processed through a lathe, a drill and a polisher. Each job has a route through machines of the kind illustrated in Fig. 3.1. The processing times or duration of each machining operation will also be specified. The machines on which the operations take place have limited capacities and a pattern of operation corresponding to the shifts of the operators, so that with several

44

hundred such operations currently going on in a factory a manager faces a large congestion problem in which queues will build up on the machines and bottlenecks will occur. The task of the planner is to forestall these bottlenecks by determining in advance the best sequences in which the operations should be processed through the machines so that the machine capacities are fully employed and the orders are completed by their due dates. The proposed schedule will specify the timetable of operations which are to take place on each machine at each shift over the planning period. This is a large and difficult problem. We will examine how far optimization techniques can be used to help with the solution of these sequencing problems.

The practical implementation of calculated sequences may present difficulties. If the operation durations on the machines are short and there is a new job on a machine every half-hour or so it may be impossible to implement detailed sequences. Also errors may arise in the data used to calculate the sequences: the actual durations on the machines may differ from the theoreti-

Fig. 3.1

cal processing time. Some of these difficulties can be overcome by relaxing control of the details of the schedule and only controlling the times at which the jobs are inlet into the factory system. We would have to have some idea how long it would take for a job to get through the factory, but provided this was known with reasonable accuracy the control problem would be greatly simplified.

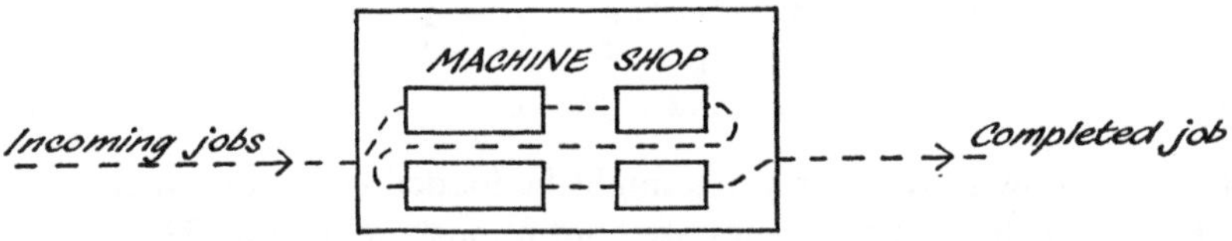

Another type of short-term planning problem arises in the efficient cutting of material. There are many industries where material is produced or stocked in certain sizes and it must be cut to different customers' requirements. Examples are the cutting of fabric to make clothes, and the slitting of paper reels, where the pieces are to be cut so as to minimize the total wastage. It is similar to a jigsaw puzzle. These cutting problems are more difficult to solve than the machine sequencing problems although they have a similar character. In the case of machine sequencing, we wish to control the start times of the operations over a period of time whereas in stock cutting we need to control the positions of the cut pieces over the length of the reel: a dimension of space replaces a dimension of time.

3.2 Optimization problems in machine sequencing

We have already met some machine sequencing problems in Chapters 8 and 9 where branch and bound procedures were used to solve the so-called three-machine scheduling problem and other simple cases. Let us recall the character of these sequencing problems by a simple example.

Example 3.1

Suppose there are three jobs each with three operations to be scheduled through three machines in turn in different orders. The machine sequences and the duration of each operation are given in Table 3.1. Fig. 3.2 shows

Table 3.1

	First operation		Second operation		Third operation	
	Machine	Duration	Machine	Duration	Machine	Duration
Job 1	1	4	2	2	3	3
Job 2	2	2	3	6	1	1
Job 3	1	3	3	4	2	3

in bar-chart form how these operations might be loaded on to the three machines, assuming that the machines are continuously available. The numbers in the rectangles correspond to the jobs. This schedule has been arranged by placing job 1's operations on the machines first at their earliest possible times followed by placing job 2's operations, followed by placing job 3's operations. Each successive set of operations regards the previous operations as irrevocably fixed.

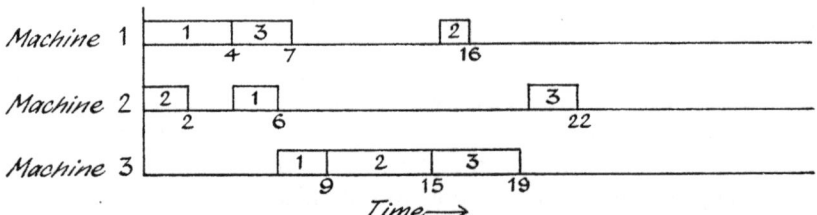

Fig. 3.2

This does not appear to be a very efficient schedule: a lot of spare time is lost on machine 2, for example, before the last operation is done. The loading of the operations could be re-arranged as shown in Fig. 3.3 The new schedule shortens the whole schedule—the last operation being completed at time 15 rather than time 22—and the average machine utilization is increased.

Example 3.1 illustrates the main features of sequencing problems. We have a number of jobs to be processed on a series of machines, each process being called an operation. As basic data for the problem we will denote by $r(j, k)$ the machine on which the kth operation of job j is to be processed and by

$d(j, k)$ its duration. The problem variables are the times at which the individual operations should be started and we will denote the start time of the kth operation of job j by $x(j, k)$. If there are N jobs and job j has $n(j)$ operations, the problem variables are thus

$$X = (x(1, 1), x(1, 2), ..., x(1, n(1)), x(2, 1), ..., x(2, n(2)), ..., x(N, n(N))$$

(We use indices rather than double subscripts in this chapter as indexed double subscripts become too cumbersome.) The $x(j, k)$ quantities are often constrained to be discrete valued.

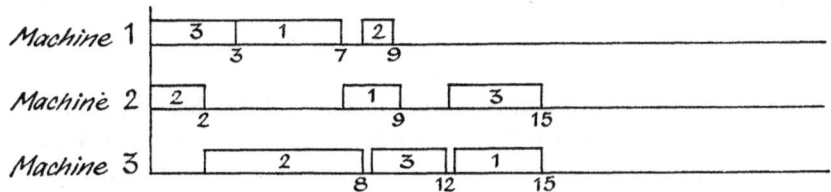

Fig. 3.3

The objective function may take a variety of forms. All the objectives aim to minimize the cost associated with the production plan. Costs may be related to the times at which jobs are completed and the efficiency with which resources are used. The most common objectives include:

(*i*)   minimizing the number of jobs which are late, or the cost associated with late jobs;

(*ii*) minimizing machine idle time;

(*iii*) minimizing the amount of work in progress;

(*iv*) minimizing the works cost through the best allocation of operations amongst alternative resources.

The objective will be expressed as a function of the operation start time variables as $F(X)$. For instance, in Example 14.1 if the objective was to minimize the average time of completion of all jobs we would minimize $F(X)$ where

$$F(X) = \tfrac{1}{3} \sum_{j=1}^{3} x(j, 3) + d(j, 3).$$

The schedule given in Fig. 3.2 would give the value 19.0 and the schedule in Fig. 3.3 would give the value 13.0. More generally we may wish to minimize the total extent to which the job completion times exceed their due dates. If $T(j)$ denotes the due time of job j, its lateness is the difference between the completion time of its last operation and the due time whenever this is positive. Therefore the sum of the latenesses is

$$\sum_{j=1}^{N} \max\,(x(j, n(j)) + d(j, n(j)) - T(j), 0).$$

The formula for idle time on say machine m is more complex. We may regard it as the non-productive time between the start of the schedule and the time at which the last operation is completed.

Two sorts of constraints arise in sequencing problems: constraints on the operations and constraints due to machine capacities. The operation timing constraints ensure that the operations follow the correct technical sequence. These constraints normally express themselves as simple linear inequalities on the start times of the operations $x(j, k)$. For example the $(k+1)$th operation of job j must not begin until after the kth operation has been completed. This would be expressed as the inequality:

$$x(j, k+1) \geqq x(j, k) + d(j, k).$$

The machine capacity constraints require that the total number of operations in any particular type of machine does not exceed the capacity. If there are $S(m)$ machines of type m available then we require that no more than $S(m)$ operations can be scheduled on machine type m at any one time. Therefore if $D(m, t)$ denotes the number of operations on machine m at time t we require that

$$D(m, t) \leqq S(m).$$

$D(m, t)$ is of course a function of the $x(j, k)$ variables but it is an awkward function to express as we shall see later.

It is rarely possible to guarantee to find the optimal solution to these sequencing problems: the functions are generally too awkward for standard non-linear optimization methods to be used, and the problems are usually too large for discrete search methods to be adopted which will reach the optimum. However, there are a few simple problems where it is possible to apply a formal optimization method and we will review these in the next section. Subsequently we will examine some extensions to more complex problems and see how heuristic and permutation procedures can be used to solve them.

*3.3 Optimal solutions to small sequencing problems

Formal optimization methods can be applied to sequencing problems of a small scale with a few jobs and a few machines. The small size of the problem means that the structures are mathematically tractable even if the contexts are seldom realistic. An important simplifying assumption which is often made is that all the jobs always go through the same sequence of machines in the same order. This problem has become known as the 'N job—M machine' problem. The computational task is very much reduced when there are at most three machines ($M \leqq 3$) as then, with the objective of minimizing total production time, there is no advantage in the jobs overtaking one another and it is only necessary to consider the order in which the jobs are loaded on to the first machine, the sequence on the second and third machines being the same.

Let us first consider N jobs going through two machines, each job requiring one operation on each of the machines A and B in that order. The processing times of job j on machines A and B are $a(j)$ and $b(j)$ respectively and a machine can process only one job at a time. The objective is to minimize the finish time of the last job to be completed on machine B. As the total processing time on the second machine is fixed as $\sum_{i=1}^{N} b(i)$ the above objective is equivalent to the minimization of the total idle time occurring on machine B. It is shown in the reference by Johnson that the jobs are optimally sequenced if we arrange them in such a way that job i preceeds job j if

$$\min\ (a(i),\ b(j)) < \min\ (a(j),\ b(i)).$$

This result is obtained by considering the effect on the accumulated idle time on machine B if jobs i and j are interchanged. The inequality provides a simple rule for sequencing jobs optimally in a two machine context with this objective function.

Example 3.2

The Table 3.2 gives example times of jobs on machines A and B for 5 jobs.

Table 3.2

job j	$a(j)$	$b(j)$
1	4	5
2	4	1
3	30	4
4	6	30
5	2	3

If the inequality rule is applied to this data the optimal sequence is determined as

5, 1, 4, 3, 2.

For example, job 5 should precede job 1 because

$\min\ (a(5),\ b(1)) = \min\ (2,\ 5) < \min\ (4,\ 3) = \min\ (a(1),\ b(5)).$

Job 4 should precede job 2 because

$\min\ (a(4),\ b(2)) = \min\ (6,\ 1) < \min\ (4,\ 30) = \min\ (a(2),\ b(4)).$

The optimal sequence gives a total production time of 47 units. If this sequence were reversed, the total time required would be 78 units, so that the optimum is a considerable improvement on some other sequences.

An alternative objective for N jobs going through two machines is to minimize the average completion times of the jobs. If $t(j)$ denotes the completion time of job j we wish to minimize

$$\frac{1}{N} \sum_{j=1}^{N} t(j).$$

This is equivalent to minimizing the sum of the completion times

$$\sum_{j=1}^{N} t(j).$$

A branch and bound procedure can be used to solve this problem along the same lines as the method used for the three-machine scheduling problem in Chapter 8. The first set of nodes correspond to all possible jobs for the first position in the sequence, the second set of nodes to all possible jobs in the the second position and so on. Any node corresponds to a subsequence of jobs which has been scheduled with the remaining jobs still to be sequenced. For instance, if there were a total of 7 jobs the branches up to the last node shown in Fig. 3.4 give the sequence 1, 4, 3, 6 and jobs 2, 5, and 7 would still have to be ordered.

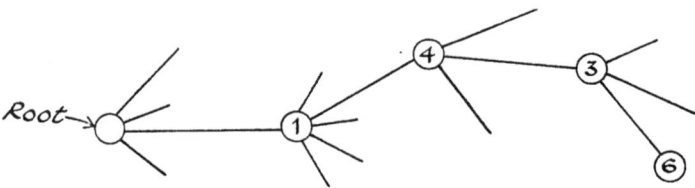

Fig. 3.4

In order to apply the branch and bound procedure we need to determine formulae for lower bounds on the average completion time corresponding to any node. As the number of jobs is fixed it is equivalent to deal with the sum of the completion times. Suppose a set I_r of r jobs has been scheduled and a set E_r of jobs has not yet been scheduled. The sum of the completion times can be written as

$$\sum_{i \in I_r} t_i + \sum_{i \in E_r} t_i.$$

The completion times for the jobs in I_r can be worked out exactly. Suppose a sequence S of jobs has already been loaded on to the machines and the earliest times at which the two machines A and B become free are $A(S)$ and $B(S)$. Then if job i is added on to the end of the sequence S, its first operation is finished on machine A at time

$$A(S) + a(i)$$

and its completion $t(i)$ on machine B is

$$t(i) = \max\left(B(S), A(S) + a(i)\right) + b(i).$$

These formulae enable us to work out the $t(i)$ values for all $i \in I_r$.

To obtain a lower bound on the excluded jobs in E_r, consider first the minimum time required on machine A. Suppose the jobs in E_r are ordered by increasing $a(i)$ value in the sequence $i(1)$, $i(2)$, ..., $i(N-r)$ where

$$a(i(1)) \leq a(i(2)) \leq a(i(3)) \leq \ldots \leq a(i(N-r))$$

and it is assumed that no delays occur on machine B. Then the completion time of job $i(p)$ is estimated as

$$\sum_{j \in I_r} a(j) + a(i(1)) + a(i(2)) + \ldots + a(i(p)) + b(i(p)).$$

This formula can be used to provide a lower bound on the sum of the completion times because of the ordering of the $a(i(p))$ quantities. The lower bound $T(E_r)$ for the excluded jobs E_r will be

$$T(E_r) = \sum_{p=1}^{N-r} [A(I_r) + ((N-r-p+1 \cdot a(i(p)) + b(i(p))))].$$

Similarly, by considering machine B as the bottleneck and ordering the jobs by increasing $b(j)$ value as $j(1)$, $j(2)$, ..., $j(N-r)$, we obtain another lower bound as

$$S(E_r) = \sum_{p=1}^{N-r} [\max(B(I_r), A(I_r) + \min_{i \in I_r} a(i)) + (N-r-p+1)b(j(p))].$$

The final lower bound in the subsequence I_r is therefore

$$L(I_r) = \sum_{i \in I_r} t_i + \max(T(E_r), S(E_r)).$$

The standard branch and bound procedure can now be applied to the sequencing problem.

Lastly, on the subject of small machine sequencing problems, it is instructive to see how the three-machine scheduling problem, which was solved in Chapter 8 by the branch and bound method, can also be formulated as a linear programming problem. It is important to notice the difficulties which arise as it is primarily these same difficulties which prevent the use of linear programming on larger sequencing problems. The three machine problem has the same form as the problems already described except that we now have a third machine C with processing time $c(j)$ for job j. We wish to minimize the time at which all the jobs are completed.

The problem variables are $x(j, 1)$, $x(j, 2)$, $x(j, 3)$ the start times of the operations of job j on machines A, B and C. The objective is to minimize the latest completion time, i.e.

minimize $\max_{1 \leq j \leq N} (x(j, 3) + c(j))$.

The operations must take place in the correct sequence, so that

$$x(j, 2) \geq x(j, 1) + a(j), \quad 1 \leq j \leq N,$$
$$x(j, 3) \geq x(j, 2) + b(j), \quad 1 \leq j \leq N.$$

C

Also we can only have one operation on a machine at any one time. This means that either job j precedes job k or job k precedes job j giving the conditional constraints for all pairs of job j, k,

either $x(j, 1) - x(k, 1) \geqq a(k)$

or $x(k, 1) - x(j, 1) \geqq a(j)$

with similar constraints for machines B and C. By introducing a zero-one variable $y(m, j, k)$ for each pair of jobs (j, k) on each machine m we can convert the conditional constraints into inequalities by the method described in Example 11.4. Also by rewriting the objective function as:

minimize t

subject to $x(j, 3) + c(j) \leqq t, \quad 1 \leqq j \leqq N$

the problem is converted into integer linear form. However, because of the very large number of variables it would be impossible to use the linear programming formulation except on the smallest problems.

Other small sequencing problems have been tackled by optimization methods: dynamic programming has been used to solve two machine problems and a 'two job—M machine' case has been formulated as a shortest route problem. But none of these methods which determine the global optimum has been successfully transferred to larger realistic situations. It seems that for these we necessarily have to resort to heuristic techniques and the specialized kind of local optima which can be obtained by permutation methods.

3.4 Complex sequencing problems

Realistic sequencing problems are much larger and more complex than the situations which have just been described. A factory may well have 30 or 40 machines and several hundred operations to be sequenced through them to provide the forward plan. The optimal solutions which have been proposed will not solve problems on this scale in a reasonable amount of computer time.

The real problems are not as neat and tidy as is suggested by the small-scale problems which have been formulated. Other factors enter the structure of the problem. Let us first consider what other characteristics may be associated with jobs in practical situations. Jobs from different customers may have different priorities associated with them and the objective function may have to ensure that jobs with high priorities finish on time. Suppose $u(j)$ denotes the priority associated with job j. Then the objective may be to minimize the sum of the latenesses weighted by their priorities. If $t(j)$ is the completion time of job j, the objective function might have the form

$$F(X) = \sum_{j=1}^{N} u(j) \cdot \max(t(j) - T(j), 0)$$

where, as before, N is the number of jobs and $T(j)$ is the due date of job j.

The timing of the jobs themselves may be interconnected by assembly commitments. These can be demonstrated by a simple example. Suppose job 3 cannot start till jobs 1 and 2 are completed. This kind of assembly structure is illustrated in Fig. 3.5, in which job 1 has four operations and jobs 2 and 3 have three. If $x(j, k)$ denotes the start time of the kth operation of job j and $d(j, k)$ is its duration, the above assembly structure would require constraints of the form

$x(3, 1) \geqq x(1, 4) + d(1, 4)$

$x(3, 1) \geqq x(2, 3) + d(2, 3).$

These constraints would have to be specified for each assembly structure.

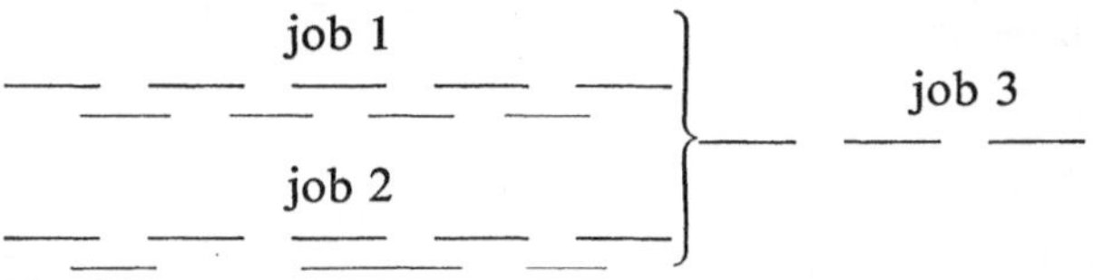

Fig. 3.5

The operations may not simply follow one another in the simple manner that has been assumed. There may be a transit time between the successive operations to allow for transporting the partially complete product between work centres. If $d'(j, k)$ denotes the transit time which must follow the kth operation of job j and $d(j, k)$ is the duration, the transit time constraint requires that the start time of the $(k+1)$th operation must satisfy the inequality

$x(j, k+1) \geqq x(j, k) + d(j, k) + d'(j, k).$

There may also be a set-up time before an operation can oe started on a machine. If this set-up time depends simply on the current operation and the machine, it can be included in the operation duration. But if a tool has to be changed on the machine, or a temperature has to be altered, or the machine has to be washed down between two products, the set-up time will depend on the predecessor operation, the current operation and the machine. Set-up considerations of this kind will lead to the same kind of difficulties for formal optimization methods as were encountered on the travelling salesman problem discussed in earlier chapters.

But the real difficulties for optimization treatment of sequencing problems arises with the resource requirement structure. First an operation may be able to go on to a number of alternative machines, not simply one unique machine. When this occurs we have to introduce a new variable $y(j, k)$ to denote which machine an operation goes on to; if a set $G(j, k)$ of machines are eligible for that operation we require that

$y(j, k) \in G(j, k).$

Secondly, an operation may not simply require one machine for its processing.

It may require a number of resources—for instance, an operator, and a die as well as a machine— and these resources may be available in varying supply over time. We will now demonstrate the complexities which arise when these last considerations are formulated as an optimization problem in terms of the start time variables $x(j, k)$.

Example 3.3

N jobs are to be scheduled through a factory, each job having a specified arrival time and consisting of a series of operations. The operations may require a number of resource types and a number of units of each resource type. The resources are in variable supply over time. The objective is to minimize the sum of the latenesses squared.

The following notation will be used:

N = Number of jobs

$A(j)$ = Arrival time of job j in the factory

$T(j)$ = Due time of job j

$n(j)$ = Number of operations for job j

$d(j, k)$ = Duration of kth operation of job j

$r(j, k, m)$ = Number of units of resource type m required by the kth operation of job j. (This quantity is assumed to be supplied at a constant rate over the duration of the operation.)

M = Number of resources

$S(t, m)$ = Number of units of resource type m which can be supplied at time t.

(All these quantities are assumed to be integer valued.)

The problem variables are the start times of the individual operations $x(j, k)$. The objective function is

$$F(X) = \sum_{j=1}^{N} \max \{x(j, n(j)) + d(j, n(j)) - T(j), 0\}^2.$$

The sequence constraints require that the jobs do not start until after their arrival times:

$$x(j, 1) \geqq A(j), \quad 1 \leqq j \leqq N;$$

and that the correct processing order is observed:

$$x(j, k+1) \geqq x(j, k) + d(j, k), \quad 1 \leqq k < n(j); \quad 1 \leqq j \leqq N.$$

The resource capacity constraints require that the operations should be timed so that the demand for each type of resource does not exceed the supply at any time. It is difficult to express this constraint. The demand for resource type m at time t by the kth operation of job j is $r(j, k, m)$ if the operation is being processed at time t, and is otherwise zero. Let us introduce new variables $\delta(j, k, t)$ to distinguish when the operation is being processed. Let $\delta(j, k, t)$ be unity if the kth operation of job j is being processed at time t and otherwise be zero. Thus the demand for resource type m by this operation at time t is

$$\delta(j, k, t) . r(j, k, m).$$

Table 3.3

job j	1	2	3	4
Start time	0	0	3	5
Duration	3	5	4	3
Resource units	2	1	1	2

As the total demand for any type of resource must not exceed the supply we require that

$$\sum_{j=1}^{N} \sum_{k=1}^{n(j)} \delta(j, k, t) . r(j, k, m) \leq s(t, m) \text{ for all } t \text{ and } 1 \leq m \leq M$$

where

$$\delta(j, k, t) = 1 \text{ for } x(j, k) \leq t < x(j, k) + d(j, k)$$
$$\delta(j, k, t) = 0 \text{ otherwise.}$$

This is a very complex form of constraint and it is worth demonstrating that it operates correctly on a simple case. Suppose we had one resource available

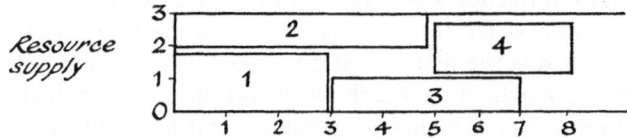

Fig. 3.6

at supply level 3 and four jobs each with one operation to be scheduled. The start times of the jobs are given in Table 3.3 together with their durations and the number of units of the resource which they require. The schedule which this produces is shown in Fig. 3.6. The $\delta(j, k, t)$ values for this schedule are shown in Table 3.4. Hence the schedule is feasible as the total resource demands given in the last row of the table are less than or equal to the supply throughout.

This completes the specification of the optimization problem. The awkward capacity constraints prevent any standard optimization procedures being applied to the problem in this form. However, in the next two sections we will consider how heuristic techniques can be used on sequencing problems almost regardless of their complexity and then see how permutation procedures can offer further improvements.

Table 3.4

t	0	1	2	3	4	5	6	7	8
$\delta(1, 1, t)$	1	1	1	0	0	0	0	0	0
$\delta(2, 1, t)$	1	1	1	1	1	0	0	0	0
$\delta(3, 1, t)$	0	0	0	1	1	1	1	0	0
$\delta(4, 1, t)$	0	0	0	0	0	1	1	1	0
$\sum_{j=1}^{3} \delta(j, 1, t) . r(j, 1, 1)$	3	3	3	2	2	3	3	2	0

3.5 Heuristic techniques

Chapter 10 considered how heuristic techniques could be regarded as methods for assigning values to a number of items by a sequential decision process, the heuristic being the rule for selecting the next item and choosing its value. This type of solution policy has been applied widely to deal with awkward machine scheduling problems. The operations effectively correspond to the items and their places in the sequences and timing are the values assigned to them.

The first stage towards applying a heuristic rule to machine sequencing is to organize the decision points. We do this by simulating the whole machine loading process, thus creating the queues which build up on the machines and ordering these by an appropriately determined priority rule. We start from the beginning of the schedule and assess the first operation of each job on the various machines. We select the machine on which an operation can be loaded earliest and choose an operation from the current queue on that machine by an appropriate criterion. When an operation has been selected and scheduled a certain amount of machine time has been committed for the duration of the selected operation so that we can reassess the times at which the next operation of the selected job can be started, and again select a job from the 'earliest' queue. We are in effect creating the queues which will occur successively through the schedule and applying a heuristic rule to determine how the queues should be sorted out.

The basic scheme can be summarized in the following steps:

(*i*)   Consider the next operations of all jobs.

(*ii*) Determine their possible start times in view of the available resources.

(*iii*) Choose a machine for which operations can be started at the earliest possible time.

(*iv*) Select one of the operations from the queue on the machine by a priority rule and load it at its earliest possible time.

This scheme is completely general and can be applied to very complex sequencing problems including set-ups, multiple resources per operation, and resources with variable supply such as the problem described in Example 3.3. The formal specification of this simulation scheme can be quite complex even in comparatively unsophisticated contexts. We will express the routine for the case where machines only process one operation at a time and an operation only requires one machine. We use the notation defined in Example 3.3, except that we consider only one resource per operation. Suppose that at any given iteration $(n'(j) - 1)$ operations of job j have already been scheduled and that $E(m)$ is the time at which the last operation on machine m was finished. Then the earliest time $e(j, n'(j))$ at which the next operation of job j could start is

$$e(j, n'(j)) = \max \left[x(j, n'(j) - 1) + d(j, n'(j) - 1), E(r(j, n'(j))) \right]$$

where $r(j, n'(j))$ is the machine required by operation $(j, n'(j))$. We now choose an operation $(j^*, n'(j^*))$ for which $e(j, n'(j))$ is a minimum over all j and include in the current subset, say S, all operations $(j, n(j))$ for which

$$e(j, n'(j)) = e(j^*, n'(j^*))$$

and

$$r(j, n'(j)) = r(j^*, n'(j^*)).$$

Only at this point do we apply a heuristic rule: we select from S the job j which minimizes some cost or priority measure $f(j)$:

$$f(j) = \min_{j \in S} f(j)$$

and then assign a start time to its next operation as the earliest possible, i.e.

$$x(j, n'(j)) = e(j^*, n'(j^*)).$$

The priority measure will be based on the objective criterion. We have to estimate from the characteristics of the operations in the queue on the selected machine which is the operation to choose to minimize the objective function. This is not easy. We are trying to make a short-term decision about a particular queue on a particular machine so as to achieve an objective referring to the total schedule. The following rules are those most commonly advocated in the research investigations which have been reported.

First come first served

This is a familiar scheme: we select the job which has been waiting longest from the queue. It is a reasonable priority rule for keeping the jobs moving but it takes no account of the differing importance of the jobs in the queue:

nor does it consider resource utilization. To apply this rule we choose $f(j)$ as the time at which the previous operation of the job was finished.

$$f(j) = x(j, n'(j)-1)+d(j, n'(j)-1).$$

The shortest operation rule

The shortest operation discipline is to select the operation from the queue which has the least duration. The case for applying this rule is that it will keep the other jobs in the queue waiting for as short a time as possible and promote resource utilization. On the other hand it will take no account of the due dates of the jobs, and operations with long durations will tend to be postponed indefinitely. The cost measure to be minimized is

$$f(j) = d(j, n'(j)).$$

Least slack

The least slack rule is a sensible heuristic for due date considerations. We measure the slack associated with a job as the time remaining between the due date and the total remaining processing time. We then select the job with the least slack. The slack associated with job j is measured as

$$f(j) = T(j)-e(j, n'(j))- \sum_{k=n'(j)}^{n(j)} d(j, k).$$

Least slack per operation

Another version of the least slack rule is to divide the total slack by the number of remaining operations and use this as a measure of slack per operation. The object of this adjustment is to favour jobs with many operations which may be held up more frequently in subsequent queues. On the other hand, once a job is late, the minimization of $f(j)$ after the slack has been divided by the number of operations is going to work in the other direction favouring jobs with few operations. The least slack per operation is measured as

$$f(j) = \left[T(j)-e(j, n'(j))- \sum_{k=n'(j)}^{n(j)} d(j, k) \right]/(n(j)-n'(j)+1).$$

Example 3.4

Suppose that the first operations of four jobs 1, 2, 3 and 4 have been scheduled as illustrated in Fig. 3.7 and that the second operations of jobs

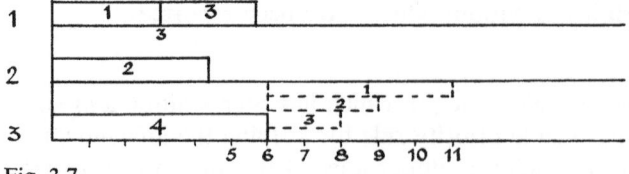

Fig. 3.7

1, 2 and 3 are now to be processed on machine 3, the durations of the operations are as shown in the diagram by the lengths of the dotted lines, with the second operations of jobs 1, 2 and 3 having durations 5, 3 and 2 respectively. If the first come first served heuristic is used the next operation to be selected for machine 3 will belong to job 1 as it clearly has been waiting longest. However, if the shortest operation rule is adopted the next operation will belong to job 3 as its processing time is least.

Very occasionally it is possible to prove that a heuristic rule will determine an optimal schedule. We will illustrate one such case. Suppose that there are N jobs waiting to be processed on a single machine and that the objective is to minimize the average completion time. Let $d(j)$ denote the duration of job j on the machine. Now it seems sensible to order the jobs by the shortest operation discipline, and in fact this will determine the optimal ordering. For convenience the indices of the jobs can be arranged so that

$$d(j) \leqslant d(j+1) \text{ for } j = 1, ..., N-1$$

giving the shortest operation rule as the ordering 1, 2, ..., N. To demonstrate that this is the optimum, suppose that the jobs were ordered in some sequence other than 1, 2, ..., N so that the kth job was the first to be out of the order 1, 2, ..., N. Then the sequence would be

$$1, 2, ..., k-1, p_k, p_{k+1}, ..., p_{k+2}, ..., p_{k+j}, k, ..., p_N,$$

where job k occupies the $(k+j+1)$th position. If job k is now moved to the kth position, the completion time for job k is reduced by an amount

$$d(p_k) + d(p_{k+1}) + ... + d(p_{k+j}).$$

The completion times of jobs $p_k, p_{k+1}, ..., p_{k+j}$ are increased by an amount $d(k)$. As the completion times of the other jobs are unaffected the net decrease in total completion time is

$$(d(p_k) - d(k)) + (d(p_{k+1}) - d(k)) + ... + (d(p_{k+j}) - d(k))$$

which is positive as

$$d(k) \leq d(p_i) \text{ for } p_i > k.$$

Thus the total completion time is reduced by making this transfer (unless all the durations happen to be equal) and hence the average completion time will be reduced. By making a series of such transfers the minimum value of the objective function will be obtained corresponding to the shortest operation sequence. The shortest operation discipline will also minimize the average number of jobs in progress and the average waiting time for single machine systems.

3.6 Permutation procedures for sequencing problems

Permutation procedures can readily be applied to sequencing problems and they can deal with fairly complex types of problems. It is natural to think of a sequence of operations on a machine as a permutation of elements and it is not difficult to visualize how we could define locally optimal permutations in terms of exchanges on the elements. In Chapter 9 a two-machine scheduling problem was formulated as a permutation problem and its local optima defined by adjacent exchanges. In practice we might well define the local optimum in terms of single shift exchanges when a computer can be used to calculate it. We will see how a permutation procedure can be applied to the multi-resource problem of Example 3.3.

Example 3.5

To formulate the problem of Example 3·3 we first need to pose the sequencing task as a permutation problem. Suppose we arrange all the operations in a list and denote the ith operation in the list by (j_i, k_i) signifying that it is the k_ith operation of job j_i. Then if there are $T = \sum_{j=1}^{N} n(j)$ operations, the list forms a permutation of T elements as

$$[P] = [(j_1, k_1), (j_2, k_2), ..., (j_T, k_T)].$$

This permutation is quite separate from any sequence on the machines at this stage.

We now need a means of transforming this permutation into a set of operation start times. To do this the operations are taken sequentially from the permutation and loaded on to the available resources one at a time. First we select operation (j_1, k_1) and load it at its earliest possible start time. Next operation (j_2, k_2) is selected and resources are allocated to it without affecting operation (j_1, k_1). Then (j_3, k_3) is taken and so on. The start time of operation (j_i, k_i) is determined after fixing the values of

$$x(j_1, k_1), x(j_2, k_2), ..., x(j_{i-1}, k_{i-1}).$$

The formula for determining the $x(j_i, k_i)$ quantities is straightforward but rather lengthy to write down. We need to record the current supply levels remaining for the various resources and ensure that sufficient supply is available over the interval $(x(j_i, k_i), x(j_i, k_i) + d(j_i, k_i))$. Also we must ensure that the operation sequence constraints are satisfied. We will summarize the means of determining the start times from the permutation by the relation

$$[X] = H[P].$$

For example, suppose an operation required 1 unit of resource type 1 and 1 unit of resource type 2 over a duration of 4 units and the previous operations had used resources as shown by the shaded areas in Fig. 3.8 where the resources are supplied at constant level 2. The scheme of loading the operation at the earliest possible time will determine the start time at 7. If the operation duration was halved to 2 units it could have been started at time 3 as there is a unit of each resource to spare over the interval (3.5). By using this scheme repeatedly we are effectively transforming the permutation into start times, and it is shown in the references that there exists a globally optimal permutation $[P^*]$ which, when transformed by this scheme, will yield the global optimum $\{X^*\}$ of the original problem stated in start time variables in Example 3.3

The objective function value associated with a permutation $[P]$ is

$$F(X) = F\{H[P]\}$$

where $F(X)$ is the lateness criterion function of Example 3.3. We can thus evaluate the total cost associated with a permutation. We must also restrict

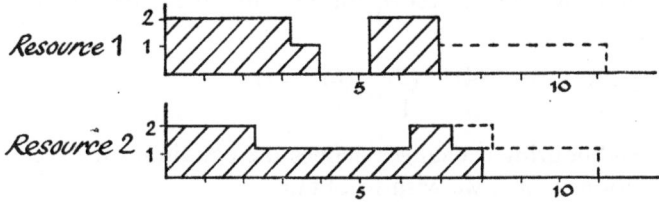

Fig. 3.8

the permutations so that the operations are in the correct sequence: the position in the permutation of the kth operation of job j must be greater than the position of the $(k-1)$th operation of job j, as otherwise the transformation may determine an infeasible schedule. We thus have a feasible set of permutations G and our optimization problem is to

minimize $F\{H[P]\}$

subject to $P \in G$.

It should be noted that it is easy to keep within the set of feasible permutations as the sequence constraints are easy to apply. The advantage of transforming the problem into the permutation form is that it gets rid of the awkward machine capacity constraints which made the problem of Example 3.3 so intractable in its original form.

The next task is to determine a suitable set of exchanges for defining the local optimum. Clearly in practical problems there will be a large number of elements in the permutation and therefore we must choose a relatively small set of exchanges. We will consider the set of exchanges corresponding to a special class of single shifts. Taking any operation (j_i, k_i) we will consider shifting it to a new position in the permutation immediately beyond its nearest operations in the permutation which use common resources with (j_i, k_i). Clearly there is no point in shifting it to a position short of this as this will have no effect. If the operations in positions a and b are the nearest operations in the permutation which use common resources with (j_i, k_i), then we consider moving (j_i, k_i) to the positions $(a-1)$ and $(b+1)$ as illustrated in Fig. 3.9.

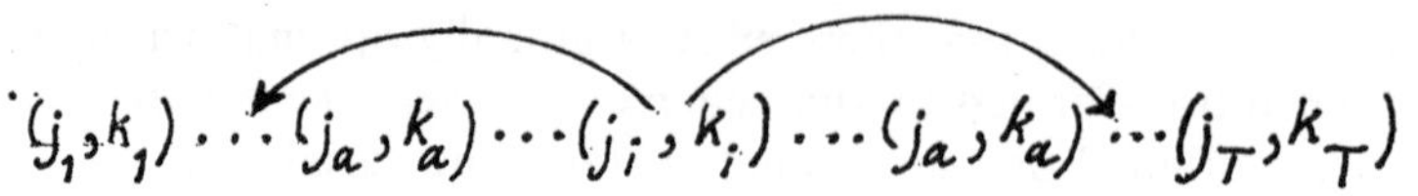

$$(j_1,k_1)\dots(j_a,k_a)\dots(j_i,k_i)\dots(j_a,k_a)\dots(j_T,k_T)$$

Fig. 3.9

We will refer to the operations in positions a and b as the resource neighbours of operation (j_i, k_i).

For example, if there were five operations numbered 1 to 7 within a permutation requiring resource numbers 1, 2 and 3 as indicated, the neighbours of operation 4 are operations 2 and 7 and the neighbours of operation 5 are operations 1 and 6 as these are the nearest operations on either side which require the same resources.

Operations	(1)	(2)	(3)	(4)	(5)	(6)	(7)
Resources types	1	3	2	3	1	1	3

The total set of such neighbouring exchanges defines the conditions for the locally optimal permutations which we wish to obtain.

To calculate a locally optimal permutation an initial permutation would be constructed and subsequently the exchange procedure would be applied. The initial permutation can be built up by choosing one operation at a time for the successive positions in the permutation. At the tth iteration we choose an operation for position t. As the objectives we are considering concentrate on lateness we choose the operation from amongst the remaining set $J^{(t)}$ of unscheduled operations as the operation for which the rate of increase in cost would be largest if it were subject to further delay. Therefore the operation selected is the next operation of job $j^{(t)}$ if the value of

$$\max (T^{(t)}(j) - T(j), 0)$$

is a maximum for $j = j^{(t)}$, and $j \in J^{(t)}$ where $T^{(t)}(j)$ is the current estimate of the completion time of job j based on the remaining processing time and the typical delays which are occurring in the queues. Having obtained an initial permutation, we can then apply the exchanges to the operations in the standard way until a locally optimum solution has been obtained.

Both the heuristic methods and permutation procedures have been applied to the problem posed in Example 3.3 and some of the results will be reproduced here. The data consisted of up to 30 jobs with between 3 and 10 operations per job. The jobs arrived randomly over the interval (0, 60), and the operation durations were drawn from a negative exponential distribution with average 2·5. The due dates were set to the arrival times plus the total processing time multiplied by a factor which varied uniformly in the interval 1·2 to 2·2. There were 10 resource types each having a constant supply level 4. The operations required up to 3 resource types and up to 3 units of each resource type.

Table 3.5 compares the average job lateness which was achieved on 5 cases using the permutation procedure with heuristic procedures listed in Section 3.5. Table 3.6 records the results of 5 other cases where the maximum job lateness is used as the criterion.

Table 3 .5

Average job lateness

Case	Initial permutation	Locally optimal permutation	First come first served	Shortest operation	Least slack	Least slack per operation
1	8·7	3·0	10·0	7·0	8·1	8·6
2	9·2	4·5	14·1	9·2	9·1	10·5
3	13·6	3·4	21·8	11·8	13·4	11·2
4	24·2	9·4	39·8	27·0	29·5	32·0
5	5·1	0	6·6	4·4	5·6	6·4

Table 3.6

Maximum job lateness

Case	Initial permutation	Locally optimal permutation	First come first served	Shortest operation	Least slack	Least slack per operation
1	0	0	62	34	12	14
2	2	2	2	4	4	4
3	0	0	15	12	8	8
4	9	0	33	40	59	59
5	0	0	10	11	10	10

It will be seen from the tables that the permutation procedure offers significant improvements over the heuristic scheduling methods with which it is compared. Furthermore, the results reached by the permutation procedure are consistently good whereas the best heuristic technique varies from case to case. Similar inconsistencies were found in a set of heuristic rules studied by Gere. Part of the weakness of the one-pass heuristic sequencing methods seems to lie in their failure to bring predecessor operations forward in time close to the final operation of the job whenever it may be completed, thereby freeing resources for other final operations which may thus be finished earlier. The neighbouring shift exchanges achieve this adjustment, and, as a by-product, this will lead to reductions in work-in-progress.

Permutation procedures can also be applied to sequencing problems in which we have a choice about the machine on which an operation is placed. We will outline this briefly in Example 3.6.

Example 3.6

Suppose we had 14 jobs each with one operation to be performed on three possible machines. The machines can process one job at a time and there are set-ups between the jobs. Fig. 3.10 shows how the jobs might be sequenced where the numbers in the rectangles indicate the jobs, and the gaps between the jobs indicate set-up times. The three sequences on the three machines can be expressed as three permutations

$[P_1] = [1, 4, 11, 14]$

$[P_2] = [3, 6, 2, 13, 12]$

$[P_3] = [7, 5, 8, 9, 10]$

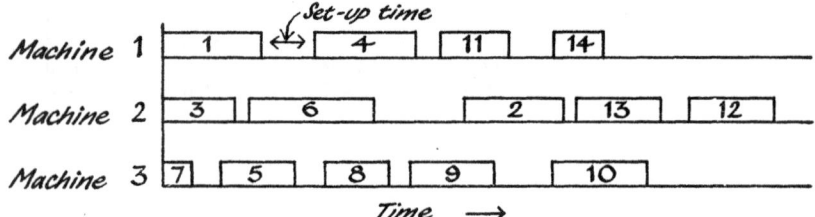

Fig. 3.10

To determine the start times of the operations from the permutations we simply take each sequence in turn and allocate the operations at their earliest start times. Suppose we were attempting to find sequences which minimized the total set-up time on all three machines subject to the total time required on any one machine not exceeding a certain limit. Then we could perform block shift exchanges moving groups of consecutive jobs with low inter-job set-up times to new positions either on the same machine or on to new positions on other machines. At each exchange we would reduce the objective function value until the local optimum had been obtained.

3.7 Inlet controlled scheduling

One of the major difficulties of applying machine sequencing procedures is to get the detailed plan implemented on the workshop floor. In the case of sequencing to minimize the costs due to job lateness it is possible to adopt a different approach in which we simply try to control the start times of the first operations of individual jobs. Whatever the start times of the jobs, we assume that having entered the factory floor they are sequenced through the machines in a fixed fashion, such as the first come first served approach. We therefore refer to this approach as inlet controlled scheduling as we only control the inlet times of the jobs.

If $x(j)$ denotes the start time of the jth job, then we can evaluate the completion times $t(j)$ by simulating the factory system using the fixed rules and the given job start times. The simulation will thus determine relationships

$$t(i) = g(x(1), x(2), ..., x(N))$$

as clearly the completion time simply depends on the times at which all the jobs are inlet into the factory system although the actual function g is not analytically known. The calculation of the $t(i)$ values may be a lengthy computation. The objective function of minimizing total lateness

$$F(X) = \sum_{i=1}^{N} \max (t(i) - T(i), 0)$$

(where $T(i)$ is the due time of job i) will thus be determined from any specification of the $x(i)$ values. Assuming that the jobs have arrival times $A(j)$ we may express the optimization problem as

minimize $F(X)$

subject to $x(j) \geqq A(j)$.

We would use a direct search approach on this problem starting from an intelligently chosen initial point based on a consideration of the due dates. The aim would be to show that by postponing the inlet times of certain jobs till some point after their arrival times the total congestion would be reduced and thus all the jobs, including those starting later may meet their due times more effectively.

Although this method appears costly in computation terms, its implementation would be remarkably straight forward, as the planning office rather than the machine operators could be totally responsible for implementing the solution.

3.8 Stock cutting to minimize wastage

Stock cutting problems can be likened to machine sequencing problems as both are concerned with efficient resource utilization. The general stock-cutting problem is to cut a number of demanded pieces of material from the reels or plates of stocked material so as to minimize the wastage. Here we will simply consider the cutting of rectangular shapes from reels of material, although more general shapes occur in practice such as the cutting of cloth for tailoring, possibly with pattern considerations to be taken into account, or three dimensional problems where the cutting takes place from solids. We will also assume that the rectangles to be cut have a fixed orientation, with a specific edge lying parallel to the edge of the reel. The problem therefore has the appearance illustrated in Fig. 3.11 in which six rectangles have been placed on a reel and the shaded area is wastage.

This problem can be formulated quite simply. Let the reel have width B and let the N rectangles have dimensions (a_i, b_i) for $i = 1$ to N where $b_i \leq B$ and the side of length b_i lies at right angles to the length of the reel. Taking the bottom left-hand corner of the reel as the origin, the problem variables may be denoted by (x_i, y_i), $i = 1$ to N, identifying the coordinates of the bottom left-hand corner of the ith rectangle. If we consider the reel to be used up to the point where the final cut is made, the total wastage to be minimized is

$$B. \quad \max_{1 \le i \le N} (x_i + a_i) - \sum_{i=1}^{N} a_i b_i.$$

If we wish to minimize the total wastage, we can equivalently minimize $\max_{1 \le i \le N} (x_i + a_i)$, as the remaining terms are constants. We must satisfy the capacity constraints ensuring that the rectangles all fit within the boundary of the reel and they do not overlap. These are difficult to express algebraically.

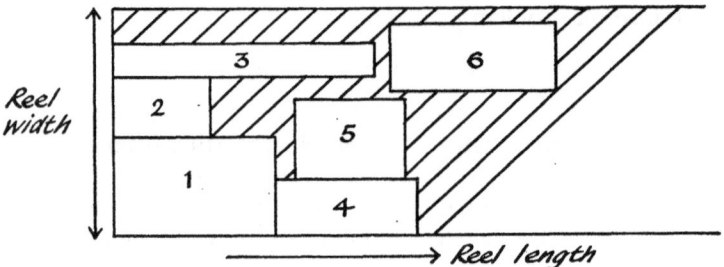

Fig. 3.11

One approach to this problem is to devise a simple rule for laying out the positions of the rectangles. Suppose we arrange the rectangles in a sequence, we can load the rectangles in turn on to the reel placing them so that their x_i coordinate is as small as possible and, where there is a choice, selecting the y_i coordinate also as small as possible. This procedure is illustrated in the following example.

Example 3.7

Ten shapes are cut by the heuristic procedure from a reel of width 6 ft. The ten shapes have dimensions given in Table 3.7. If we use the cutting scheme

Table 3.7

i	1	2	3	4	5	6	7	8	9	10
b_i	2	1	2	1	2	2	3	2	5	1
a_i	5	3	4	1	3	7	6	3	1	2

the coordinates of the bottom left-hand corners of the rectangles are specified as shown in Table 3.8. The positions are illustrated in Fig. 3.12, and the total wastage is 17 square units. This is not a very efficient rule. It is possible to re-place the rectangles in a feasible pattern so that only 5 square units are wasted and this re-arrangement can be found visually quite simply. However, it is much more difficult to implement by computer. Pattern recognition and manipulation of this kind is unsusceptible to automatic computation. Although the general two-dimensional stock-cutting problem has also been

Table 3.8

i	1	2	3	4	5	6	7	8	9	10
x_i	0	0	0	0	4	4	7	11	14	1
y_i	0	2	3	5	2	4	0	3	0	5

formulated as a linear programming problem the computational details are very complex and no practical applications have been recorded. Linear programming is much more suitable for the trim problem described in the next section which has been used profitably in practice.

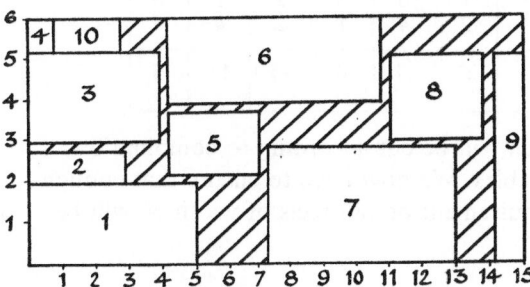

Fig. 3.12

3.9 The paper-trim problem

The paper-trim problem is a particular application of wastage minimization which is used successfully in the paper industry. Paper mills produce paper in reels of given width, the width depending on the particular manufacturing machine. Customers require reels of various widths, and the reels of standard width must therefore be cut, and often some wastage at the edge will occur. The manufacturer wishes to make the trim loss as small as possible. We can use linear programming to solve this problem; it is illustrated in the following example.

Example 3.8

The standard reels are of width 215 inches and the orders given in Table 3.9 have been received. First we assess the various ways in which a standard reel of 215 inches can be partitioned into units of the demanded widths. The ten possible ways are shown in Table 3.10, each possibility being referred to as a combination. For example, the first combination considers 3 widths of

64 inches leaving a trim waste of 23 inches, the second combination slices the standard reel into 2 widths of 64 inches and 1 width of 60 inches the remaining waste being 27 inches, etc.

Table 3.9

Width in inches	Number of reels ordered
64	180
60	90
35	90

Table 3.10

Combination i	1	2	3	4	5	6	7	8	9	10
Widths 64	3	2	2	1	1	1	0	0	0	0
60	0	1	0	2	1	0	3	2	1	0
35	0	0	2	0	2	4	1	2	4	6
Trim waste	23	27	17	31	21	11	0	25	15	5

Denote the number of reels which will be cut according to combination i by x_i; these are the problem variables. We now have to ensure that enough of each width is produced. The requirement of 180 reels of width 64 will be satisfied if:

$$3x_1 + 2x_2 + 2x_3 + x_4 + x_5 + x_6 \geq 180$$

and similarly, for the other orders, we need:

$$x_2 + 2x_4 + x_5 + 3x_7 + 2x_8 + x_9 \geq 90$$
$$2x_3 + 2x_5 + 4x_6 + x_7 + 2x_8 + 4x_9 + 6x_{10} \geq 90.$$

By using inequalities we overcome the difficulties which might arise by requiring the x_i values to be integral. The slack variables have a special significance. If we produce any surplus above the order then it is essentially further waste. Therefore if x_{11}, x_{12}, x_{13} are the slack variables attached to the constraints the wastage associated with these quantities is $64x_{11}$, $60x_{12}$ and $35x_{13}$.

The complete system of equations when slack and artificial variables x_{14}, x_{15}, x_{16} have been added is

$$
\begin{aligned}
3x_1 + 2x_2 + 2x_3 + x_4 + x_5 + x_6 \qquad\qquad\quad -x_{11} \qquad\quad +x_{14} \qquad\qquad &= 180 \\
x_2 \quad\quad +2x_4 + x_5 \quad +3x_7 + 2x_8 + x_9 \qquad\quad -x_{12} \qquad\quad +x_{15} \quad &= 90 \\
2x_3 \quad\quad +2x_5 + 4x_6 + x_7 + 2x_8 + 4x_9 + 6x_{10} \qquad\quad -x_{13} \qquad\quad +x_{16} &= 90
\end{aligned}
$$

The total trim wastage to be minimized is the sum of the trim wastes on the slicing patterns and any surplus on the orders:

$$23x_1 + 27x_2 + 17x_3 + 31x_4 + 21x_5 + 11x_6 + 25x_8 + 15x_9 + 5x_{10} + 64x_{11} + 60x_{12} + 35x_{13}$$

which is a linear function. Furthermore, the x_i values must all be positive. We have thus formulated the problem as a standard linear program.

The solution to this problem is $x_1 = 60$, $x_7 = 30$, $x_{10} = 10$ and otherwise $x_i = 0$. It is illustrated in the diagram of Fig. 3.13.

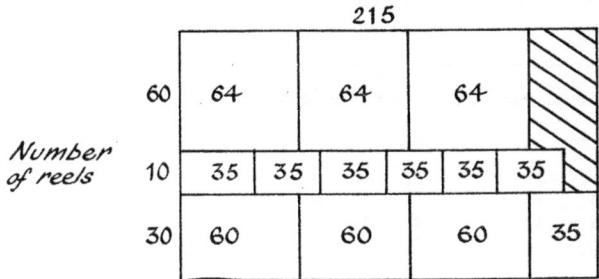

Fig. 3.13

REFERENCES

Bulkin, M. H., Colley, J. C. and Steinhoff, H. W. 1966. Load forecasting, priority sequencing and simulation in a job shop control system. *Man. Sci.*, **13** (2), 29.

Conway, R. W. and Maxwell, W. L., 1961. Network despatching by the shortest operation discipline. *Opr. Res.*, **10**, 51.

Gere, W. S. 1966. Heuristics in job shop scheduling. *Man. Sci.*, **13** (3), 167.

Giffler, B. and Thompson, G. L. 1960. Algorithms for solving production scheduling problems. *Opr. Res.*, **8**, 487.

Gilmore, P. C. and Gomory, R. E. 1965. Multistage cutting stock problems of two and more dimensions. *Opns Res.*, **13**, 94.

Heller, J. and Logeman, G. 1960. An algorithm for constructing feasible schedules and computing their schedule times. *Man. Sci.*, **8**, 168.

Johnson, S. 1954. Optimal two and three stage schedules with set-up time included. *Naval Research Logistics Quarterly*, **1**, 61.

Kelley, J. E. 1961. Critical path planning and scheduling: mathematical basis. *Opr. Res.*, **9**, 296.

Manne, A. S. 1960. On the job shop scheduling problem. *Opr. Res.*, **8**, 219.

Mellor, P. 1966. Job shop scheduling—a review. *Opr. Res. Quat.*

Muth, J. F. and Thompson, G. L. 1963. *Industrial Scheduling*. Prentice-Hall, New York.

Nicholson, T. A. J. and Pullen, R. D. 1968. A Permutation procedure for job-shop scheduling. *Computer J.*

Rowe, A. J. 1960. Toward a theory of scheduling. *J. Ind. Eng.*

Sisson, J. 1960. Sequencing Theory. Chapter 7 in *Progress in Operation Research*. Ed. R. L. Ackoff, J. Wiley, New York.

Wagner, H. M. 1959. An integer linear-programming model for machine shop scheduling. *Naval Logistics Quarterly*, **6**, 131.

Vajda, S. 1958. *Readings in Mathematical Programming*. Pitman, London.

Exercises on Chapter 3

1 In Section 3.3 a formula was devised for calculating a lower bound on the average completion times of all jobs for the 2-machine N-job problem. Using the data in Table 3.2 calculate this lower bound when jobs 1 and 4 have been scheduled in that order.

2 Apply the heuristic techniques of the shortest operation rule and the least slack rule to the following sequencing problem. There are four jobs each with three operations to be performed on three machines. All jobs arrive in the factory at time 0. Each machine can process one operation at a time and the machines are always available. The data for the jobs is as follows:

Table 3.11

Job	Due time	Machine sequence	Operation durations
1	14	1, 2, 3	3, 2, 5
2	20	3, 1, 2	5, 1, 3
3	18	2, 1, 3	4, 4, 4
4	15	1, 3, 2	2, 6, 2

Compare the maximum job lateness achieved by the two rules and also the total machine idle time, where idle time is the non-productive time on a machine between the first and last operations processed on it.

3 The permutation procedure used in Example 3.5 converts a permutation of the operations into a set of job start times by taking each operation in turn and loading it on to the available resources at the earliest possible times without disturbing the positions of the preceding operations in the permutation. Use this scheme to determine the start times of the single operation jobs with durations and resource requirements as shown in Table 3.12 where the permutation is 1, 2, 3, 4, 5, 6 and the resources of types 1 and 2 are available at constant supply level 3. Illustrate the resource usage in a diagram.

Table 3.12

Job number		1	2	3	4	5	6
Operation duration		3	2	3	4	1	2
Resource requirements	type 1	1	0	1	1	2	2
	type 2	2	1	1	3	2	0

4 Three jobs are to be processed through two machines *A* and *B* twice, so that each job goes through the processes in the order *A, B, A, B*. A machine can only process one job at a time and the machines are continuously available. The times which the jobs take on the two machines are halved on the second pass and the times on the first pass are in hours as follows:

Table 3.13

	A	B
Job 1	6	6
Job 2	4	2
Job 3	2	4

All jobs can start at time 0 and it is assumed that a job can start immediately on the next process on completion of the previous process. Apply the shortest operation discipline to determine the start times of the operations.

If the jobs all had due times of 19, and the objective was to minimize maximum lateness, formulate this problem as a permutation problem. Write down a permutation which would offer an improvement over the schedule obtained by the shortest operation rule.

5 At the beginning of a period a paper factory has outstanding orders for N products. $W(i)$ tons of product i are required by a due date $T(i)$, otherwise a penalty cost will be incurred of $C(i)$ units per hour late. The factory has one production line, and it requires $r(i)$ hours to produce 1 ton of product i. The total order for a product is made in a continuous batch. The machine is reset between products, $S(i, j)$ being the resetting time if product j follows product i. The cost of running the machine is b units per hour and it is assumed that it is never worth while for the production line to be idle. The objective is to minimize the total costs of the schedule. Formulate the optimization problem as a permutation problem, and show how the order completion times are determined. Given the following product, cost and set-up data, show that the permutation [3, 2, 4, 1] is optimal with respect to adjacent exchanges.

Table 3.14

$b = 1$,

i	1	2	3	4
$T(i)$	15	13	12	24
$C(i)$	2	2	2	2
$W(i)$	2	3	2	5
$r(i)$	2·5	3	2	2

Table 3.15

$S(i, j) =$

i \ j	1	2	3	4
1	0	3	6	2
2	5	0	4	2
3	6	1	0	3
4	1	3	7	0

6 Formulate the linear programming problem for the following paper trim problem with reel width 82 inches.

Table 3.16

Width in inches	Number of reels required
58	60
26	85
24	85
23	50

7 In some cutting problems where rectangular pieces are being cut from a reel of material or a long slab of metal the mechanics of cutting requires that any cut should go from edge to edge of the material. The reel is therefore cut into smaller and smaller rectangles. These are called guillotine cuts. A heuristic scheme for cutting rectangles from a reel where guillotine cuts are necessary would be to take the widest rectangle (and where there is a choice the longest) and place it on the lower edge of the material, cut across the reel and then cut down the length of the reel to form the rectangle. The spare rectangle could then be used by treating it as a small reel and cutting it up for the widest rectangle that will fit into it, and then cutting the next piece of spare and so on. When no more rectangles could be cut from the spares we would return to the original reel and cut the next largest rectangle from this. We would continue in this way until all the required rectangles had been cut. Apply this scheme to the data of Table 3.7. Assume that the reel has a width of 6 units, that the rectangle orientation is fixed, for instance due to a grain in the material, and that we always cut from the largest spare rectangle first.

*8 Critical path planning is concerned with the timing of interdependent activities or operations in a project so as to minimize the completion time of the project. The interconnections between the operations can be expressed by a network of the form illustrated in Fig. 3.14. The operations are the

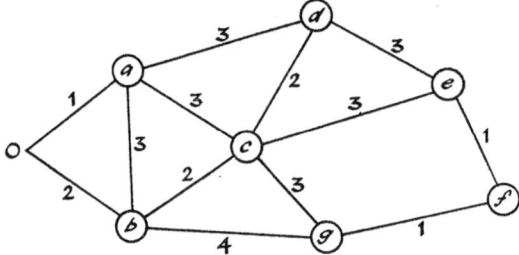

Fig. 3.14

connections of the network with the durations marked on them, and each operation is identified uniquely by the pair of numbers forming the ends of the connection. Each operation has an arrow marked on it to show the direction of increasing time and the nodes are called the start and end nodes for the operation. Thus the operation numbered (2, 3) cannot start until operation (1, 2) has been completed, and operation (3, 5) cannot start until operations (1, 3) and (2, 3) have been completed. When limited resources have to be allocated to these operations, the operations are in effect being sequenced on to the available supply and a good heuristic rule is to order any queues of operations by the least slack rule. The slack here is more difficult to calculate. It is defined as the difference between the latest time at which the operation can start in order to complete the project in its minimum time and the earliest time at which the operation can possibly start. Show in general how to determine the slack associated with each operation by making a dynamic programming calculation on the network assuming that the operations have known durations. Use the scheme to determine the slack on operation (3, 5).

*9 A single system consists of N jobs to be processed through one machine. Job j has due date $T(j)$ and duration $d(j)$. The jobs are to be sequenced so as to minimize the maximum lateness. A good heuristic rule to achieve this objective is to sequence the jobs in order of their due dates. Show that this heuristic will in fact determine the optimum schedule. Also demonstrate by an example that this heuristic would not necessarily lead to the optimum result if the maximum deviation from the due time was being minimized regardless of whether the job is early or late.

*10 Show that in the 2-machine scheduling problem with N jobs going through two machines A and B in the same order there is never any advantage in the jobs overtaking one another where the objective of minimizing total production time.

4 Stock control and production scheduling

4.1 Decisions in stock control and production scheduling

The previous two chapters have studied very short-term problems; the optimal settings for chemical plants for which the adjustments in the controls are made every day or even every hour, and the best sequences of jobs on machines calculated every week. The stock control and production scheduling problems which are studied in this chapter are rather longer term, arising over months rather than weeks. We will be concerned with the setting of production targets and typical stock levels rather than planning the detailed way in which the usage of machinery can be optimized to meet these targets. Typical problems include the determination of how much raw material should be held in stock, how stockholding of finished goods should be balanced against new production runs, how much of each product should be made and how frequently. Although these decisions are longer term than process control and machine sequencing they may properly be defined as medium-term issues as the really long-term factors such as maximum productive and storage capacity and the structure of the market will be assumed to remain fixed. (Some of these long-term problems such as plant renewal and investment planning will be studied in later chapters.)

Two new factors are introduced with these medium-term problems. First they form a link-up between the producing unit, the factory, and the market demanding its products. Process control and machine sequencing ask no questions about the market: they are entirely internal to the production unit, optimizing the use of resources for given requirements. All the problems in this chapter will make statements about a demand pattern and make assumptions about a market structure. Furthermore, the involvement with a market introduces questions of uncertainty about market behaviour. This is the second new factor: the need to treat stochastic situations. The probabilistic features create some of the major difficulties for solving stock control problems, but they are essential for realistic solutions to practical applications.

The main sources of stock control and production scheduling problems are contained in the model shown in Fig. 4.1. The production unit draws on raw materials stocks and also produces goods for finished product stocks which then go on to meet a market demand. Stocks of raw materials and finished products are held to off-set the set-up costs of making new purchase orders for raw materials or making new runs of a product. They offer the potential economies of long production runs and bulk purchases and act as buffers against fluctuations in demand or purchasing conditions. The stock levels must not be excessive as storage costs money; the storage costs may reflect the lost opportunities for the capital tied up in stocks, the cost of

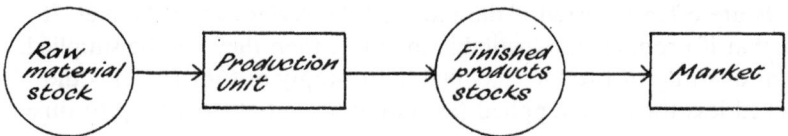

Fig. 4.1

space for storage and the risk of obsolescence. It is a question of getting the right balance.

The production unit itself may have several optimization problems. It may be able to operate at various levels each at a different cost, and its fixed capacity may be able to produce a variety of products over time. When the selling price of a product line varies over time as well as the production costs there may be significant returns to be gained by careful planning of the production targets for each product and the selection of which products to manufacture in each production cycle.

The principal variables of the model of Fig. 4.1 are defined for each product at each period under consideration. If there are N products and T periods, the following variables are defined for $i = 1$ to N and $t = 1$ to T:

$x_i(t)$ = production level for product i in period $(t, t+1)$

$s_i(t)$ = quantity of product i in stock over period $(t, t+1)$

$q_i(t)$ = quantity of product i demanded at time point t.

(It is assumed that the products are supplied at discrete moments in time.)

The $x_i(t)$ and $s_i(t)$ quantities are the problem variables; the $q_i(t)$ quantities are given data although they often have to be expressed by a statistical distribution. For simplicity we have not defined variables for raw material stock-holding. The problems of raw material stocks are very similar to those of finished goods stocks. The issues in the ordering of materials for production are the same as the issues in the ordering of production for product stocks: they just occur at a different stage in the line.

The objective in these problems is to minimize the total costs (or maximize the profits when there is a revenue in the problem). Component costs arise from three sources: the production levels $x_i(t)$, stockholding costs, penalty costs for shortages and variations in the selling prices. The production costs from three sources: the production levels $x_i(t)$, stockholding costs, penalty costs for shortages and variations in the selling prices. The production costs may depend on the product, the time interval, and what was produced in the previous period. The stockholding costs usually depend on the type of product, the quantity of product being stored and the period for which it is held. If $c_i(t)$ denotes the cost of storing 1 unit of product i in period $(t, t+1)$, the storage costs are

$$\sum_{t=0}^{T} \sum_{i=1}^{N} c_i(t) \cdot s_i(t)$$

Penalty costs are often incurred if the required deliveries cannot be met. It is assumed that if products are available in stock, then they will be supplied. But the total quantity of product i available to supply at time $(t+1)$ is the total produced less the total supplied (assuming no initial stock), up to time $(t+1)$, i.e.

$$\max\left[\sum_{k=0}^{t} x_i(k) - \sum_{k=0}^{t} q_i(k), \, 0 \right].$$

If this quantity is less than $q_i(t+1)$ there will be a shortage and a penalty cost may be incurred. Finally the selling prices of the products may vary over time. The stock control and production scheduling problem is to maximize the revenue less the costs

(revenue from sales) − (production costs) − (storage costs) − (shortage costs).

(In many of the problems we will simply be minimizing the costs.)

There may be various restrictions on the $x_i(t)$ and $s_i(t)$ values. Clearly $x_i(t)$ and $s_i(t)$ are interrelated as we can only store what is produced. But there may also be a varying lag time between the instant of production and the time of entry into stocks which complicates the relationship. This is particularly true for the ordering of materials and their receipt. The production levels $x_i(t)$ may be restricted to a few fixed settings. The total capacity for storage may also be limited and this can become an important factor when a number of products are being produced all competing for the same production and storage facilities.

Some of the simpler cases formed the first established studies in operations research: rather like linear programming on the techniques side, stock control procedures were the first application of operations research to receive widespread recognition. We will group the different situations into a number of models all of which fit within the general model which has just been constructed. The models are classified by whether they deal with one or many products, whether the demand is known or probabilistic, and whether the environment is static or changing over time. We will study the following four groups of models which are listed in order of increasing complexity.

Static deterministic single-product models

For these problems the demand $q_i(t)$ is assumed to be known exactly and not to vary over time. A single product is considered. As the environment is static, costs and selling prices do not alter and can therefore be ignored.

Static probabilistic single-product models

Here, the demand has random features although the statistical characteristics are steady over time. Again a single product is considered and we have to balance the cost of probable and possible shortages against excessive safety stocks.

Dynamic deterministic single-product models

In this group the demand and prices and costs are allowed to vary over time in a known fashion. A single product is considered.

Static deterministic multi-product models

The models in this class consider the effects of inter-product interactions either at the production or storage stage.

4.2 Static deterministic single-product models

For these models we have the simplest situation of a production unit producing to replenish stocks for a known and constant demand. Production takes place on the receipt of an order to replenish stocks and the production will enter stock either as it is being produced, or possibly as a batch after some delay. The demand is met from stocks as requested and, if the demand cannot be met on time, it may be backlogged and delivered late. The costs of making a new production run, the costs of storage, and the costs of late deliveries are known.

As all quantities are steady over time the amount to produce (or order) over time will assume a steady pattern. As there is a set-up cost associated with production we can expect the production to take place in batches at regular intervals. Therefore, rather than formulate the problem in terms of the variables, $x(t)$ (the production rate at time t), the problem can be assessed in terms of the intervals between production runs and their timing in relation to demand. As the demand occurs at a constant rate the amount to produce in the batch will also be governed by the interval between production runs assuming that demand must be met.

Example 4.1

A retailer holds a stock of goods to meet a demand occurring at a constant rate r per unit time. The retailer can obtain new supplies instantly at a set-up cost of c_1. The cost of holding 1 unit in stock for 1 unit of time is c_2. It is assumed that deliveries are never late.

If x denotes the initial stock level after a batch arrives we can graph the level of stocks as they are depleted as shown in Fig. 4.2. The time taken for a batch of size x to be used up is $\frac{x}{r}$, whereupon a new batch will be ordered and the process repeated.

We wish to determine the batch size x so as to minimize the total costs over a long period of time. This is equivalent to minimizing the average costs or costs per unit time per cycle. The costs per cycle consist of a set-up cost

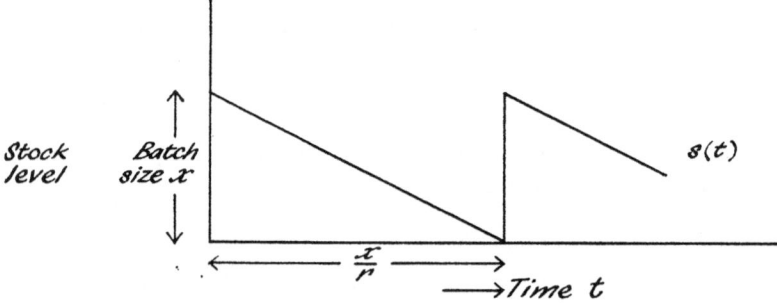

Fig. 4.2

and the stock-holding costs. If $s(t)$ denotes the stock on hand at time t, the costs per cycle are

$$c_1 + c_2 \int_0^{x/r} s(t)dt$$

$$= c_1 + c_2 \int_0^{x/r} (x - rt)dt$$

$$= c_1 + \frac{c_2}{2} \cdot \frac{x^2}{r}.$$

The quantity $\frac{x^2}{2r}$ can be obtained alternatively by considering the area of the triangle in Fig. 4.2. The cost per unit time $F(x)$ is obtained by dividing the costs per cycle by $\frac{x}{r}$:

$$F(x) = r\frac{c_1}{x} + \frac{c_2}{2} \cdot x.$$

This can be minimized by differentiation to give

$$\frac{dF}{dx} = 0, \text{ when } x = \sqrt{\left(\frac{2rc_1}{c_2}\right)}.$$

Thus the most economic batch size is $\sqrt{\left(\frac{2rc_1}{c_2}\right)}$ and the batches are re-ordered

every $\sqrt{\left(\frac{2c_1}{rc_2}\right)}$ time units.

This very simple example can be generalized to include costs for shortages as illustrated in the next example.

Example 4.2

Suppose that we have the same conditions as occurred in the previous example but that we now have a cost c_3 per unit per unit time delay on delivery of the product. Although we can clearly order the product frequently enough to avoid delays, it may be worth paying the costs of some delayed deliveries so as to reduce storage costs if the storage costs are comparatively high. The graph of storage $s(t)$ would appear as shown in Fig. 4.3. The negative portion of the curve shows the extent of the shortages.

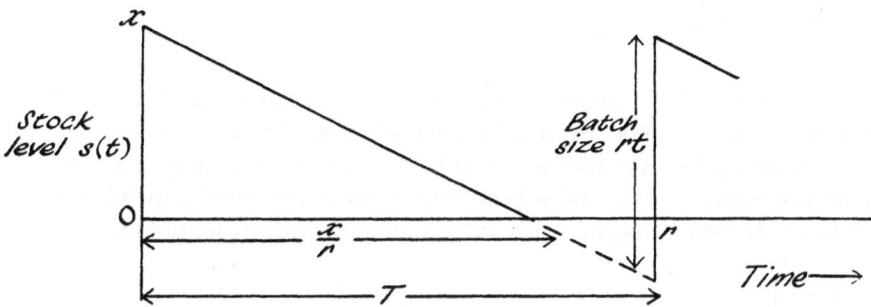

Fig. 4.3

Let x be the initial stock level immediately after a new order arrives in stock and after backlogged demand has been satisfied. Then shortages begin after time $\frac{x}{r}$. Let us also denote by T, the time between re-ordering. The batch size will be rT of which $(rT-x)$ will be delivered immediately.

The total costs over the interval $(0, T)$ are the set-up costs c_1 plus stockholding costs,

$$c_2 \int_0^{\frac{x}{r}} s(t)dt = c_2 \int_0^{\frac{x}{r}} (x-rt)dt = \tfrac{1}{2}c_2 \frac{x^2}{r},$$

plus the shortage costs

$$c_3 \int_0^{T-\frac{x}{r}} s(t)dt = c_3 \int_0^{T-\frac{x}{r}} rtdt = \tfrac{1}{2}c_3 \frac{(rT-x)^2}{r}.$$

Thus the total costs per unit time $F(x, T)$ are now expressed as

$$F(x, T) = \frac{c_1}{T} + \frac{c_2 x^2}{2rT} + c_3 \frac{(rT-x)^2}{2rT}.$$

There are no constraints on the problem variables x and T (except the requirement that they should be positive) so the minimum can be found by differentiating with respect to x and T to obtain

$$\frac{\partial F}{\partial x} = \frac{1}{T}\left[\frac{c_2 x}{r} - \frac{c_3(rT-x)}{r}\right]$$

$$\frac{\partial F}{\partial T} = -\frac{1}{T^2}\left[c_1 + \frac{c_2 x^2}{2r} + \frac{c_3(rT-x)^2}{2r}\right] + \frac{c_3(rT-x)}{T}.$$

Equating $\dfrac{\partial F}{\partial x}$ and $\dfrac{\partial F}{\partial r}$ to zero gives two equations, which can be solved for x and T to give the results:

$$x = \sqrt{\left\{\frac{2rc_3 c_1}{c_2(c_2+c_3)}\right\}},$$

$$T = \sqrt{\left\{\frac{2c_1(c_2+c_3)}{rc_2 c_3}\right\}}.$$

These formula specify the optimal initial stock levels after a re-order and the intervals between re-orders. The total shortage before a re-order is $(rT-x)$.

It may be noted how the formula is to be adjusted if no shortages are to occur. In this case $c_3 = \infty$, and if both numerator and denominator of both formulae are divided by c_3, and this value substituted for c_3 we obtain

$$T = \sqrt{\left(\frac{2c_1}{rc_2}\right)}$$

$$x = \sqrt{\left(\frac{2rc_1}{c_2}\right)}$$

which, as would be expected, are the same formulae derived in Example 4.1. If there are no set-up costs, i.e. $c_1 = 0$, then $x = 0$ and $T = 0$ showing that ordering or 'production should take place as demand requires it eliminating both storage and shortage costs. Finally, if there are no storage costs, i.e. $c_2 = 0$, the formulae give $x = \infty$ and $T = \infty$ showing that the production should take place only once and an infinite amount should be stocked.

As an illustration of these formulae, suppose a manufacturer supplies 25 engines per day. The cost of holding a completed engine in stock is £16 per month, and there is a penalty of £10 per engine per day late. Production takes place in batches, and the set-up costs are £10,000.

Taking 1 day as the unit of time, £1 as the unit cost, and 1 month as 30 days, the data gives $c_1 = 10,000$, $c_2 = 16/30$, $c_3 = 10$ and $r = 25$. The formulae give the optimal interval between batches as $T = 40$ days and the initial stock level after production as 943 engines. The total batch size is 1000.

In the exercises at the end of the chapter we consider a single product deterministic model in which the production rate is limited and the batch is fed into storage at a fixed rate. This is a common condition of practical situations.

4.3 Probabilistic demand: the re-order level and periodic review systems

The most unrealistic aspect of the models described in the previous section is the assumption that demand will remain at a constant rate. The demand for goods varies, particularly consumer goods, and this must be taken into account. A second factor which was ignored was the possibility of delay between the time at which the re-order is made and the time at which the goods actually arrive in stock. This is called a lead time and it often occurs in practice. It is these two factors in combination which give rise to the difficulties. Let us first consider them individually.

Suppose we have a probabilistic demand but no lead time. Then we could order a new batch as soon as the stocks fell to the zero level. Furthermore, we could calculate the appropriate batch size and re-ordering interval in exactly the same way as described in Section 15.2 using average characteristics of demand. If p(i) denotes the probability of a demand of i units per unit time the average demand rate per unit time is r where

$$r = \sum_{i=0}^{\infty} i \cdot p(i).$$

Thereafter the same formulae apply as were used before. Equally, if demand was at a constant rate and there was a lead time, the new batch could be re-ordered at a time sufficiently far in advance to allow the stocks to dwindle to zero just before the batch arrives. If the demand rate was r and the lead time L we would make the re-order when there were rL units left in stock.

However with a positive lead time and a probabilistic demand we may unintentionally run out of stock and need to hold a reservoir of safety stocks. The aim of this section is to discuss the correct setting for these safety stock levels. It is assumed that the lead time is known and that the statistical distribution of demand does not change over time. Furthermore, the batch size and the average re-order interval are assumed to be determined by considering the average rate of demand and using the same formulae as were discussed in the previous section.

There are two main methods for treating the safety stocks problem. The first method, known as the re-order level system, aims to determine an optimal stock level at which to re-order a new batch. The intervals between ordering will vary with this system depending on the particular character of demand. It is also known as the two-bin system in which one bin of stock holds an amount of stock equal to the re-order level quantity, and the other major bin holds the remainder: re-ordering takes place as soon as all the stock has been drawn from the major bin. It is clearly a very simple system to implement. The second method of treatment, known as the periodic review system, examines the stock position at regular intervals and brings the stock up to some appropriate maximum level. As demand varies, the quantity in the order will vary. This system is particularly useful when a range of stock items is ordered from a single supplier, as it may lead to quantity discounts. The details of the two systems are examined in the next two examples.

Example 4.3. The re-order level system

The demand for production from a depot varies randomly but the statistical distribution of the demand rate is known with average value r. When stock

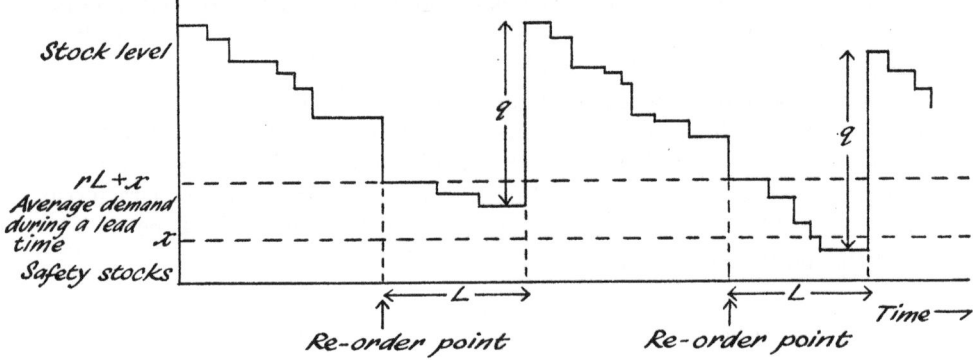

Fig. 4.4

runs out it is re-ordered but there is a lead time L between the instant of re-order and the time at which the products reach the depot. The quantity q to re-order is determined by considering the average demand. The decision to re-order is made when the stock level is equal to the average requirements rL during a lead time plus a safety stock x. The diagram of Fig. 4.4 shows how the stock level will vary over time. In the first lead time on the diagram the demand does not call upon safety stocks, whereas in the second period we do require the safety stocks. The batch size on each occasion is the same, and we assume that unsatisfied demand is backlogged.

We wish to determine the level of safety stocks x which minimize the costs associated with storage and shortage. The storage costs are denoted by c_1 per unit per unit time stored. It is assumed that any shortages are backlogged but that the shortage costs are c_2 per unit short regardless of the actual time delay on delivery. If the safety stocks are too large, unnecessarily high storage costs will be incurred, whereas, if x is too small, the penalty costs will be comparatively too large. The optimization problem is to find the minimum cost balance between these two sources of cost connected with the safety stocks.

The total costs will be measured as an average cost per average cycle time q/r where q is the batch size and r the average demand rate. The average stockholding costs per cycle of holding safety stocks x is

$$c_1 x \cdot q/r$$

as the average demand will be q and the safety stocks will be held over the full cycle. The average shortage costs depend on the demand structure. If $p(y)$ denotes the probability of y units being demanded during a lead time, the expected shortage costs will be

$$c_2(1 \cdot p(x+rL+1)+2 \cdot p(x+rL+2)+3 \cdot p(x+rL+3)+...).$$

If we denote the cumulative probability that demand is equal to or greater than y during the period L by P_y,

$$P_y = \sum_{i=y}^{\infty} p(i),$$

the shortage costs may be expressed as

$$c_2(P_{x+rL+1}+P_{x+rL+2}+P_{x+rL+3}+...).$$

The objective function $F(x)$ expressing the total expected costs per cycle is now

$$F(x) = c_1 x \frac{q}{r} + c_2(P_{x+rL+1}+P_{x+rL+2}+P_{x+rL+3}+...).$$

We now wish to minimize $F(x)$. This can be done by examining the discrete gradient of $F(x)$. Starting from some large value of x we would reduce x so long as the following condition held:

$$F(x+1)-F(x)>0.$$

(This is the discrete equivalent of the statement $\frac{dF}{dx}>0$ showing that the direction of descent is pointing in the direction of decreasing x.) Now

$$F(x+1)-F(x) = c_1(x+1)\frac{q}{r} + c_2(P_{x+rL+2}+P_{x+rL+3}+...)$$

$$-c_1 x\frac{q}{r} - c_2(P_{x+rL+1}+P_{x+rL+2}+...)$$

$$= c_1 \frac{q}{r} - c_2 P_{x+rL+1}.$$

We therefore require the smallest value of x such that

$$F(x+1)>F(x)$$

$$P_{x+rL+1} < \frac{c_1}{c_2} \cdot \frac{q}{r}.$$

D

This condition can be used to determine the optimum value of the re-order level x.

As an illustration of the use of the formula, suppose the demand for a warehouse item follows a Poisson distribution with a mean of 6 per week and the lead time is 1 week and the batch size is 24. The storage cost of holding an item over one week is 1 shilling and the shortage cost is £4 per unit. The probability of a demand y or more can be tabulated as shown in Table 4.1.

Table 4.1

y	0	1	2	3	4	5	6	7	8	9	10	11	12
P_y	1·00	1·00	0·98	0·94	0·85	0·71	0·55	0·39	0·35	0·15	0·08	0·04	0·02

The average demand during a lead time rL is 6 since the lead time is 1 week, and the smallest value of y such that

$$P_y < \frac{0·2}{4} = 0·05$$

is at $y = 11$, giving

$x + rL + 1 = 11$.

Hence $x = 4$.

The safety stock is 4 units.

Example 4.4. The periodic review system

In this system the stock level is inspected at fixed intervals and a quantity is ordered to bring the stock on hand plus that on order up to some fixed

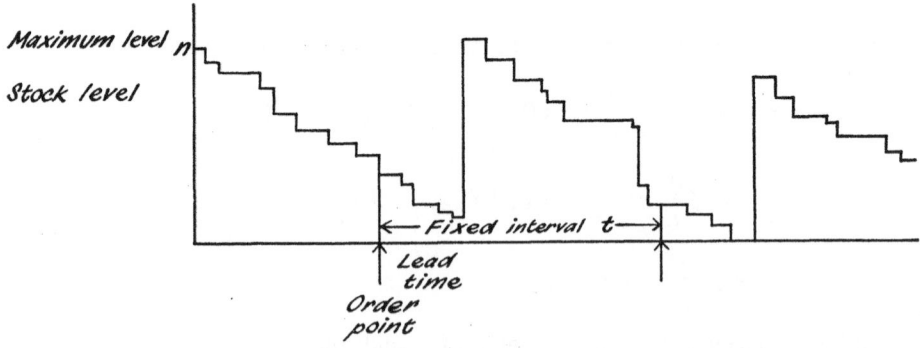

Fig. 4.5

maximum level. The variable demand in the lead time waiting for the stock replenishment will mean that the stock level does not always come up to the same level. The graph of stock level will take the form shown in Fig. 4.5.

Suppose n is the maximum stock level which we wish to reach after each intake and let t be the interval between inspections. Suppose, for instance that the lead time was 1 week, the interval between inspections was 4 weeks, and the average demand was 100 units per week with a safety stock of 50 units. Then every fourth week an order for 400 units would be made when the stock level was 150 units to bring the stock, on the average, up to 450 units one week later. Again the order has to be large enough to take care of any excess demand which occurs in the lead time and balance this against excess storage cost. The detailed formulation of this problem is studied as an exercise.

4.4 Dynamic models: warehousing, production smoothing, reservoir control

Both the previous sections have discussed essentially static situations in which the environment does not change over time although some of its characteristics may be described by statistical distributions. However, the nature of demand does vary as time goes on, and if these variations can be predicted it may be best to determine varying production and stockholding policies which recognize this variation. Furthermore, in the entirely static situation prices and costs were ignored yet they may be significant factors in stockholding policies if they vary, for instance due to seasonal influences. In this section we will formulate a number of examples of production scheduling and stockholding problems where the conditions change over time. We begin with a purchasing problem where the product's price and cost varies over a number of periods. Next we examine two production scheduling problems: in the first, the variation in demand requires that the extra overtime costs needed to produce large batches have to be balanced against storage costs. The second problem considers how to smoothe production levels against a varying demand, when there are costs associated with changing the level of demand. Lastly we will briefly examine how dynamic programming can be used to formulate an optimal control policy for a reservoir where the stock is the current level in the reservoir and the source of supply depends on probabilistic rainfall patterns.

Example 4.5. A warehousing problem

The warehousing problem is concerned with the intelligent purchasing and selling of a product in a fluctuating market. Suppose a warehouse has fixed capacity and an initial stock of a product which is subject to known seasonal price and cost variations; we wish to determine the optimal pattern of purchasing, storage and sales. The demand r_i in each period is known; if it is not met no penalty is incurred. Let B be the storage capacity of the warehouse, and v be the initial stock of the product. Also let c_i be the cost per unit, p_i be the selling price per unit in period i, there being a total of N periods. There is a fixed storage cost of c per unit per period.

The problem variables are the quantities which should be purchased and sold. Let x_i, y_i and s_i $(i = 1, ..., N)$ be the quantities which are purchased, sold and stored respectively at the beginning, end and middle of period i respectively. It is thus assumed that the product has to be held for at least one period. The objective function is the sum of the profits in each period

$$F(X, Y, S) = \sum_{j=1}^{N} (p_j y_j - c_j x_j - c s_j).$$

This is to be maximized subject to three restrictions. First the storage in any period must not exceed the maximum capacity B,

$$s_i \leq B.$$

Secondly, the amount which is sold in any period cannot exceed the total stock on hand or the demand,

$$y_i \leq s_i$$
$$y_i \leq r_i.$$

The stock on hand in the ith period is related to the stock in the $(i-1)$th period as

$$s_i = s_{i-1} + x_i - y_{i-1}$$
$$s_1 = v + x_1.$$

Finally, the amounts bought, sold and stocked must be non-negative:

$$x_i, y_i, s_i \geq 0, \quad i = 1, 2, ..., N.$$

Clearly this is a linear programming problem and can be solved by the simplex method.

Example 4.6. Balancing regular time, overtime and storage

A manufacturer is planning production over the next n months and in the ith month must supply r_i items of a given commodity. The items can be produced in regular operator's time up to a ceiling of B_i units per month or in overtime up to a ceiling of B_i' units per month. The cost of producing one item in the ith month is c_i on regular time and c_i' on overtime. The storage cost per unit is d per month and the maximum quantity which can be stored is D. The variation in cost with time, and also the capacity restrictions, might make it more economical to produce in advance of the period when the items are actually needed. It is assumed that the demands must be met.

Let x_i and x_i' denote the number of items produced in the ith month in regular and overtime respectively. Also let s_i denote the number of items stored over during the ith month. These are the problem variables. s_i can be expressed in terms of x_i and x_i' as

$$s_i = x_1 + x_1' + x_2 + x_2' + ... + x_{i-1} + x_{i-1}' - r_1 - r_2 - ... - r_{i-1}$$

Then, by the principle of optimality,

$$f_k(y) = \min_{x_k \geq r_k} (g_k(x_k - r_k) + h_k(x_k - y) + f_{k+1}(x_k)).$$

This recurrence relation is evaluated for $k = N, N-1, N-2, ..., 1$ and for a range of suitable values of y. A calculation on numerical data is provided as an exercise. Dynamic programming is well suited to this kind of problem where decisions for adjacent periods are connected, but otherwise the separate decisions are independent.

Example 4.8   *A reservoir control system*

Hydro-electric schemes which operate from the storage of water in large reservoirs are faced with the problem of optimizing the control of water discharge. The dynamic and probabilistic feature of the problem is the highly seasonal pattern of water input from river flows into the reservoir. Given the pattern of demand for electric energy, it is required to minimize the cost of meeting these demand requirements. We will outline the way in which the problem can be formulated.

The data and variables of the problem relate to each of the N periods. Let s_i denote the water stock in the reservoir at the beginning of period i; let y_i denote the river flow into the reservoir during period i, let x_i denote the water usage or discharge from the reservoir during this period and let r_i denote the demand for electric energy. The problem variables are the inter-related quantities s_i and x_i. The technology of the system is such that the cost incurred during period i depends on the stock level, the amount of water discharged, the river flow and the demand. We may denote this cost as

$g_i(x_i, s_i, y_i, r_i)$.

The river flow y_i can only be measured in terms of a probability distribution. Since river flows are highly correlated over time the probability density function of y_i is assumed to be of the form $p_i(y_i \mid y_{i-1})$ signifying conditional dependence on the flow in the previous period. The total expected costs therefore have the form:

$$F(x_1, x_2, ..., x_N; s_1, s_2, ..., s_N)$$
$$= \sum_{i=1}^{N} \int_{-\infty}^{+\infty} g_i(x_i, s_i, y_i, r_i) \cdot p_i(y_i \mid y_{i-1}) dy_i.$$

This is to be minimized subject to the constraints relating the stock levels at periods i and $(i+1)$, and the discharge in period i. The relation may be written as

$$s_{i+1} = h_i(s_i, x_i).$$

The problem can be solved by dynamic programming. If we let $f_i(s_i, y_{i-1})$ be the cost of an optimal policy for periods i to N starting with a stock level s_i in period i and the previous month's inflow y_{i-1}, the recurrence relations for dynamic programming relate the optimal policy over the remaining

$(N-i)$ periods to the optimal policy over the remaining $(N-i-1)$ periods. This is expressed as the recurrence relation between f_i and f_{i+1}:

$$f_i(s_i, y_{i-1}) = \min_{x_i} \int_{-\infty}^{+\infty} [g_i(x_i, s_i, y_i, r_i) + f_{i+1}(s_{i+1}, y_i)p_i(y_i \mid y_{i-1})]dy_i$$

where we take s_i and y_{i-1} as our state variables and we substitute for s_{i+1} in terms of $h_i(s_i, x_i)$ before minimizing with respect to x_i over a suitable range of values. It can be seen that at each stage i a one-dimensional minimization is required for each pair of state variable values s_i and y_{i-1}.

*4.5 Multi-product problems: production, storage and selection

The problems of scheduling and stock control become more difficult when there are interactions between the stock levels or production levels which can be obtained for the different products. A production facility which manufactures a number of different products normally has to account for a set-up time between production runs on any product line. It is therefore desirable to have long production runs to reduce the ratio of set-up time to production time. On the other hand, long production runs imply large storage costs between them and therefore, from the storage point of view, short runs are desirable. We will examine the nature of this balancing problem more precisely.

Suppose that there are N products being made and we are examining the production and stock control policies over M periods. There is a known demand r_{ij} for product i, $(i = 1, N)$ in period j, $(j = 1, M)$. Let t'_i be the set-up time to start a production run for product i, and c'_i be the set-up cost. Also let t_i be the time required to produce one unit of product i and c_i be the cost of storing it over one time unit. We will assume that demand must be met, that the total productive time in the factory is T_j in period j, and that the storage costs associated with a period are measured in terms of the surplus stock at the end of the period. It is also assumed that a group of products are made in a period so that a product run is shorter than a period. The problem is to determine the amounts of each product which should be produced in each period to minimize total costs.

Let x_{ij} be the amount of product i which is produced in period j. Then the total costs may be expressed as the storage costs and set-up costs:

$$F(x_{11}, x_{12}, ..., x_{NM}) = \sum_{i=1}^{N} \sum_{j=1}^{M} c_i \left(\sum_{k=1}^{j} (x_{ij} - r_{ij}) \right) + c'_i \delta(x_{ij})$$

where

$$\delta(x_{ij}) = 1 \text{ if } x_{ij} > 0$$
$$= 0 \text{ otherwise.}$$

We must ensure that the demand is satisfied, represented by the constraint:

$$\sum_{j=1}^{k} (x_{ij} - r_{ij}) \geq 0; \quad \text{for } k = 1, ..., M.$$

Also the productive capacity must not be exceeded,

$$\sum_{i=1}^{N} t_i x_{ij} + \sum_{i=1}^{N} t'_i \delta(x_{ij}) \leq T_j.$$

All the x_{ij} variables must be non-negative.

This completes the formulation of the optimization problem. It is not easy to solve. The $\delta(x_{ij})$ quantities prevent the use of linear programming. It may well be worth using a heuristic technique to obtain an initial solution to the problem which could then be improved by steadily adjusting the x_{ij} values reducing F at each step as in a descent or direct search method.

We will examine two variations of this multi-product problem. The first problem is to decide how long a production cycle time should be assigned to a single machine producing a number of different products to meet a given demand. The second type of problem arises where we are not necessarily bound by demand but are limited by capacity. The problem is to determine the best combination of products to manufacture. It is known as the product mix problem.

Example 4.9

A firm manufactures N different products on a single machine facility. The demand for product i occurs at a constant rate r_i. The machine can make product i at rate p_i and the set-up and storage costs are c'_i and c_i as before. The products are made in a strict sequence one after the other, the quantities

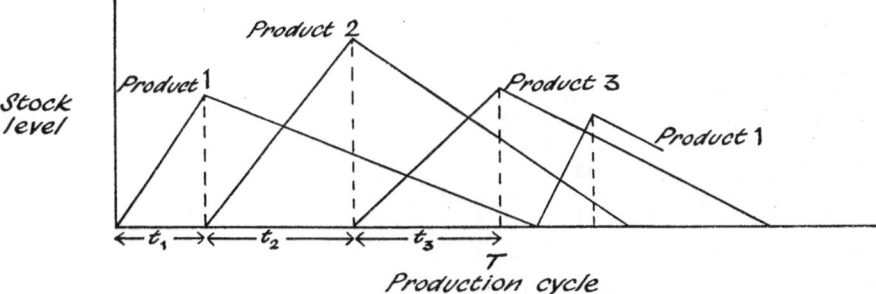

Fig. 4.6

produced being proportional to the demand rates. If the products are produced in the order 1, 2, 3, ..., N, product 1 starts to be made again when the production of product N is complete. The stock levels for three products will take the form illustrated in Fig. 4.6 in which the machine successively makes products 1, 2, 3. The problem is to determine the total production cycle time T. If T is too large stockholding costs will be too high, and if T is too small too many set-ups will occur. Within the cycle time T, a time say t_i is allocated to the production of product i. This means that the quantity produced is $p_i t_i$ and this must equal the demand, i.e.

$$p_i t_i = r_i T.$$

The total costs per cycle may be expressed as the set-up costs plus the storage costs. As the maximum storage level is $(p_i t_i - r_i t_i)$, these costs are

$$\sum_{i=1}^{N} \left\{ c_i' + T \frac{c_i}{2} (p_i - r_i) t_i \right\}.$$

Substituting for t_i as $t_i = \dfrac{r_i T}{p_i}$ in this expression and dividing by T we obtain the costs per unit time as

$$F(T) = \sum_{i=1}^{N} \left[\frac{c_i'}{T} + T \frac{c_i r_i}{2} \left(1 - \frac{r_i}{p_i} \right) \right].$$

Differentiating with respect to T and equating to zero we obtain the optimal cycle length $\overline{T}$ and the optimal lot sizes $\overline{q}_i$ of product i as

$$\overline{T} = \left\{ 2 \sum_{i=1}^{N} c_i' \Big/ \sum_{i=1}^{N} c_i r_i (1 - r_i/p_i) \right\}^{\frac{1}{2}}$$

and $\overline{q}_i = r_i \overline{T}$.

As an illustration of the formula, suppose that three products are made on a machine with the data as shown in Table 4.4.

Table 4.4

Product i	1	2	3
Demand rate r_i	4	10	5
Production rate p_i	10	30	20
Cost of set up c_i'	22	35	25
Cost of storage c_i	1	1	1

The optimum cycle time $\overline{T}$ given by the formula is

$$\overline{T} = 3 \cdot 6.$$

It should be noted that for capacity to be sufficient we require that

$$\sum_{i=1}^{N} \frac{r_i}{p_i} \le 1.$$

It should also be pointed out that we have not assumed any set-up times. If set-up times are present, and are dependent on the preceding job, it may be very advantageous to optimize the sequence within the production cycle. This type of machine sequencing optimization was discussed in Chapter 3.

Example 4.10

Suppose a firm has the potential of manufacturing N products and there is a maximum demand for r_i units of product i in a given period. However the firm is not committed to meet the requirements. The firm has M machine processes and the production of 1 unit of product i requires h_{ij} hours of time on process j. The total number of hours available on machine j is H_j over the relevant period. Product i sells at p_i per unit and the cost of using 1 unit of time for product i on machine process j is c_{ij}. The firm wishes to determine how much of each product should be produced so as to maximize profits.

Let x_i be the amount of product i to produce. Then the total profit is:

$$\sum_{i=1}^{N} \left(p_i - \sum_{j=1}^{M} c_{ij}h_{ij} \right) x_i.$$

The x_i values are constrained by non-negativity,

$$x_i \geqq 0, \quad i = 1, ..., N,$$

by the production capacity limitations

$$\sum_{i=1}^{N} h_{ij}x_i \leqq H_j, \quad j = 1, ..., M$$

and by the market limitations:

$$x_i \leqq r_i, \quad i = 1, ..., N.$$

The optimum production levels can now be determined by linear programming.

REFERENCES

Burgin, T. A. and Wild, A. R. 1967. Stock control—experience and usable theory. *Opl. Res. Quat.*, **18** (1).
Carter, L. and Johnson, R. 1966. An application of linear programming. *Scientific Business.*
Hansmann, F. 1962. *Operations Research in Production and Inventory Control.* Wiley, New York.
Krone, L. N. 1964. A note on economic lot sizes for multi-purpose equipment. *Man. Sci.* **10** (3).
Sasieni, M., Yaspan, A. and Friedman, L. 1959. *Operations Research—Methods and Problems.* Wiley, New York.

Exercises on Chapter 4

1 A simple production system consists of a production line and a storage depot. Products are required at a constant rate r, and production can take place at a rate $R>r$. There is a fixed set-up cost per production run and a known cost of storage per unit per unit time. Determine the size of production batches and the intervals between their arrivals in the depot assuming that no shortages are allowed to occur.

Apply the results to the following case. A firm has to supply 50 items per week and it can produce at a rate of 200 items per week. The cost of storing 1 item for 1 year is £10 8s. 0d. and the cost of a set-up is £375. How frequently should a production run be made and what size of batch should be manufactured.

2 If in the numerical part of Example 4.3 the storage cost increased to 10 shillings per week and the shortage cost increased to £100 per item what would the level of safety stocks be?

3 A stationer has the problem of deciding how many copies of a magazine he should stock each week. The demand varies between 20 and 30 with the probability distribution given in Table 4.5.

Table 4.5

Demand	20	21	22	23	24	25	26	27	28	29	30
Probability	0·03	0·06	0·09	0·13	0·16	0·16	0·11	0·10	0·08	0·05	0·03

The magazine costs 4 shillings for the stationer and sells at 6 shillings. How many magazines should the stationer stock?

4 In the periodic review system stock is re-ordered at fixed points in time separated by an interval of fixed length t and the amount of replenishment is such as to bring the level of stock on hand less the average amount demanded in the lead time up to a fixed level n. The lead time before the arrival of the replenishment is L. The probability of x units being demanded in a period L has a known distribution $p(x)$. The set-up cost for making an order is c_1, the cost of stock holding is c_2 per unit per unit time, and the penalty cost in the case of shortage is c_3 per unit (independent of time). Derive a formula for the total expected cost per unit time in terms of t and n, assuming that demand over a cycle decreases steadily and that in a typical cycle it begins at the value n.

5 A warehouse planner has a problem of the type described in Example 4.4 and has the following pattern of cost data c_j, selling price data p_j and demand data r_j for his products over 5 periods as shown in Table 4.6. The warehouse has a storage capacity of 200 and sets off with an initial stock of 100 units.

There is a stock holding cost of 5 units per item per period. Formulate the linear programming problem.

Table 4.6

Period j	c_j	p_j	r_j
1	20	20	160
2	22	35	180
3	18	30	140
4	25	25	120
5	40	50	170

6 A firm has made arrangements to supply a company with 210 assemblies in January, 140 in February, 180 in March, and 160 in April. Using the standard shifts at cost £20 per unit, the firm can produce only 150 assemblies per month. By working the regular shift overtime at cost £25 per unit an additional 30 assemblies can be made. Finally the firm can contract out at a cost of £30 per unit to an unlimited extent. If assemblies are not ready on time a penalty charge of £10 per unit is made per month. The manufacturer has guaranteed to produce the total requirement by the end of the period. Formulate the problem of determining the production schedule so as to minimize total costs assuming that items are never held in stock.

7 A company is planning its production schedule for the next 4 months. Monthly demands, which must be met, are as follows:

Table 4.7

Month	1	2	3	4
Demand	130	100	90	120

The cost of changing the production level from month to month is the difference between the production levels squared. Any excess production during the month does not carry over to the next month, and in fact there is a cost incurred equal to the square of the excess. Determine monthly production to minimize the cost of changing production and the cost of excess production over the whole period. Assume that the production during the month preceding the first month of the table was 120 units.

8 A firm manufactures a variety of material handling equipment, and wishes to determine what would be the optimal production levels for their products without regard to market considerations. The manufacture of a product requires three processes: fabricating, machining and assembling, and the times required (measured in standard hours) for manufacturing one unit of each of five products are given in Table 4.8. Also shown in the table are the total daily capacities in standard hours for each department, and the profit from each product line. Write down the details of the optimization problem.

Table 4.8

Product group	Hours required per item			Profit margin
	Fabricating	Machining	Assembling	
Transporting units	45	35	12	243
Lifting units	50	30	14	616
Storing units	50	30	15	893
Carrying units	175	100	75	1367
Hoisting units	1·5	13	2	63
Capacity in hours	438	1043	633	

*9 A firm has four factories each capable of producing either of two components. The components are assembled into a complete product in the ratio of three units of the first component to four units of the second component. Each factory has a working week of 80 hours. The times needed to produce components are (in hours):

	Component	
	1	2
Factory A	7	8
B	6	7
C	4	6
D	1	5

The firm wishes to maximize weekly output of the finished article without producing surplus of either component. Formulate the optimization problem in linear programming terms.

*10 A selling agency of limited capacity C takes on stocks of N types of items at fixed intervals. The value of item i is v_i and its size for storage purposes is w_i. The number of items of type i which are demanded varies according to a known distribution, $p_i(x_i)$ being the probability that x_i items of type i are demanded. Formulate the optimization problem of determining the quantities of stocks which should be accepted and state what method might be used to solve the problem.

*11 In Example 4.3 the problem of determining safety stock levels in a probabilistic environment was investigated. It was assumed that the lead time between making an order for stock and receiving the order was known and fixed. Consider briefly what would be the implications of the lead time itself being a probabilistic variable.

*12 If in the warehousing problem described in Example 4.5 the demands r_i had to be met exactly show how the problem of minimizing costs can be expressed exactly as a dynamic programming calculation.

5 Design optimization

5.1 Design optimization studies

Optimization techniques are becoming increasingly popular as an aid to the designer over the whole range of industrial activity. The general aim of a design study is to specify the make-up of a product, a machine, an instrument or even a whole system so that it possesses definite properties for performing its tasks. Traditionally the design process has been based on trial-and-error methods and high-quality designs have been achieved as experience has accumulated. However, whenever it is possible to specify an exact or even approximate relationship between the properties of the object being designed and the control variables of the design, the use of an optimization technique may mean that good designs can be obtained very much more quickly. The conditions which favour the use of an optimization technique are that there should be a comparatively large number of possible values for the design variables which would all produce a feasible solution. The techniques are particularly useful when the variables are interacting in a non-linear fashion. The optimization task is then to find the value of the design variables which produce a system which satisfies the required properties at minimum cost. Instead of simply producing a feasible design, we can now search amongst all the possible designs and actually determine the optimum one. This is the main advance which optimization techniques combined with a computer can realize in the design field.

Design studies generally lead to a non-linear optimization problem. Suppose there are N design variables denoted by x_1, x_2, ..., x_N and let there be L

equality constraints expressing exact conditions which the design must satisfy:
$$H_k(x_1, x_2, ..., x_N) = 0, \quad k = 1, ..., L.$$
Also let there be M inequality constraints which express minimum or maximum property specifications on the design:
$$G_k(x_1, x_2, ..., x_N) \leqq 0, \quad k = 1, ..., M.$$
The specification of the cost of the design $(x_1, x_2, ..., x_N)$ is denoted by $F(x_1, x_2, ..., x_N)$, and the problem is then to minimize $F(X)$

subject to $H_k(X) = 0, \quad k = 1, ..., L,$

$$G_k(X) \leqq 0, \quad k = 1, ..., M.$$

Often the functions H_k, G_k and even F may not be known exactly. Design problems and research investigations are closely related activities, so that designers are likely to encounter uncertainties about the functional forms. However, it is usually possible to guess the nature of the relationship and the details of the coefficients can be calculated by multiple regression techniques such as were described in Chapter 1, Vol. II.

Cost is not the only possible objective in a design study. An alternative design criterion is to maximize some performance factor, although one of the constraints in this case may be a cost limitation. For example, rather than minimize cost to achieve a given characteristic of a machine, we may maximize the characteristic subject to a ceiling on the costs of the final design.

In the following sections we shall investigate a wide range of applications of design optimization in product mixing, chemical plant design, machinery design, instrumentation and electronics. Unlike some of the other subjects for optimization techniques, design studies require some detailed technical knowledge of the object being designed. However, we shall attempt to outline the principles of the technologies in sufficient depth to illuminate the source of each study and demonstrate the role of optimization techniques in the field. The main achievements of the techniques will be reduced cost or improved performance, although, if the design techniques are implemented by computer program, significant savings may also be achieved in design time and effort.

It is worth noting here that design studies extend beyond technical specifications into aesthetic aspects. For example, the design of a motor car must combine mechanical factors with considerations of appearance. The lines and connections of the surfaces of the bodywork must unite both attractiveness and functional points of view. Similar commitments arise in architectural design. We shall not discuss these aesthetic factors further; they are often very difficult to quantify but are clearly an important factor in some areas of design optimization.

5.2 Design of product mixes

The first stage in the production of many products in the rubber and plastics industries is a mixing process. A number of ingredients are mixed together to form a composite material which is then processed mechanically by pressing, extrusion, etc., to form the final products. The characteristics of the final product, such as durability, are governed by the choice of the materials which go into the initial mix and their relative proportions. Usually the final product will have to satisfy a number of performance limits, but, within these limits, we will want to design the product at minimum cost. The procedure will be demonstrated on a rubber compounding design.

Example 5.1

Rubber products are made by mixing four ingredients, raw rubber, crumb (or scrap), oil and carbon black. The proportions of the ingredients determine such properties as hardness, elasticity and tensile strength in the final product. The proportions will be denoted by x_1, x_2, x_3, x_4. The relationship between the kth property say y_k and the ingredient levels can be expressed by a general quadratic relationship as:

$$y_k = a_0 + \sum_{i=1}^{4} a_i x_i + \sum_{i=1}^{4} \sum_{\substack{j=1 \\ j>i}}^{4} a_{ij} x_i x_j.$$

(The quadratic terms are necessary to take account of the interactions between the ingredients.) The values of the coefficients a_0, a_i and a_{ij} can be determined by regression techniques.

For a particular study 39 trial measurements were made to determine the relationships for three properties y_1, y_2 and y_3. The x_i values were scattered over the region $0 \leq x_i \leq 100$ using an experimental design. The regression relationships were determined as:

$$y_1 = 34 \cdot 0 + 0 \cdot 11 x_1 + 0 \cdot 06 x_2 - 0 \cdot 68 x_3 - 0 \cdot 27 x_4 + 0 \cdot 001 x_1 x_2$$

$$+ 0 \cdot 002 x_2 x_3 + 0 \cdot 001 x_2 x_4 + 0 \cdot 005 x_3^2 + 0 \cdot 003 x_3 x_4 + 0 \cdot 001 x_4^2$$

$$y_2 = 16 \cdot 7 + 0 \cdot 34 x_1 - 0 \cdot 043 x_2 - 0 \cdot 002 x_1^2 - 0 \cdot 001 x_1 x_2 - 0 \cdot 001 x_1 x_4$$

$$y_3 = 12 \cdot 4 - 0 \cdot 10 x_2 - 0 \cdot 001 x_2^2.$$

The objective function form $F(x_1, x_2, x_3, x_4)$ was also calculated by regression techniques as

$$F(x_1, x_2, x_3, x_4) = 9 \cdot 3 + 0 \cdot 006 x_1 - 0 \cdot 023 x_2 - 0 \cdot 028 x_3 + 0 \cdot 035 x_4.$$

The function $F(x_1, x_2, x_3, x_4)$ was then minimized subject to the constraints on the properties

$$25 \leq y_1 \leq 35$$

$$16 \leq y_2$$

$$12 \leq y_3$$

and on the x_i values

$25 \leqq x_1 \leqq 75$

$0 \leqq x_2 \leqq 100$

$20 \leqq x_3 \leqq 50$

$50 \leqq x_4 \leqq 100.$

Using the direct search technique the optimum mix was found to be (75·0, 22·9, 50·0, 52·2) with a cost of 9·39.

If the optimization calculation yields even a small cost reduction through finding the cheapest mix it may lead to significant annual savings as hundreds of thousands of units of the product may be made each year.

5.3 Design of chemical plant

In Chapter 2, Vol. II on chemical process control we studied how to operate chemical plants in an optimal manner. The design of chemical plants is a relate problem where we have the additional freedom of the reactor-design itself. Instead of setting controls to maximize current profits we will estimate typical operating profits for a period of years and choose the reactor design which maximizes the lifetime operating profits on the capital and running costs of the design.

The design of a chemical plant involves a detailed knowledge of the chemical process for which the plant is being designed. The resulting optimization problem is often fairly complex, involving non-linear functions. A number of studies in this field have been successfully tackled by the projected gradient method. The more sophisticated design optimization studies have attempted to optimize a whole system of components. For instance, the reference by Singer considers the optimization of a refinery model consisting of distillation columns, catalytic cracking units, polymerization and hydrogenation plants, blending facilities, storage tanks and other utilities, and the reference by Gelbin examines the optimization of a multi-stage ammonia reactor. The following example considers the design of a single reactor.

Example 5.2

A chemical reactor consisting of a single vessel is to be designed to perform the following operational requirement. Two chemical reactants are to be inlet into the vessel, and after the reaction has taken place for a certain time, called the holding time, the products having commercial value will be separated off from the remainder and the vessel will be recharged. The reaction may be assisted by additional heat input or a catalyst.

The overall objective of the design study is to maximize the total profit. The profit consists of the return from the operation of the reactor over the lifetime of the plant less the initial capital cost. The operational return depends on the increased value of the output compared with the inputs, and their rate of flow through the reactor. If V is the volume of the reactor in cubic feet and h is the time in hours for which the reactants are held in the reactor (the holding time) the flow rate is $\dfrac{V}{h}$ cu ft/hr. If the reactor is assumed to last n years, and the cost of the input reactants is c per cu ft (including catalyst and extra heat) and the price of the output product is c' per cu ft the total operational profit is

$$n.\frac{V}{h}(c'-c).24\times 365 \text{ (assuming full time operation)}.$$

It is presumed that the principal design variables of the reactor are its volume V and the temperature of operation T so that the capital cost will be expressed as a function $C(V, T)$. Thus the total design optimization problem will be to maximize the total profit from the investment as a function of V and T,

$$F(V, T) = n.\frac{V}{h}.(c'-c).24\times 365 - C(V, T).$$

We will now discuss the form of the capital costs $C(V, T)$ in more detail for a specific situation. These costs comprise the costs of the materials for the reactor casing, the insulation system, the costs of a mixer motor and the costs of construction. Let us assume that the reactor consists of a cylindrical shell made of steel of length 1·75 times its diameter capped by semi-ellipsoidal ends with a mixer motor inserted as shown in Fig. 5.1. Thus if d is the

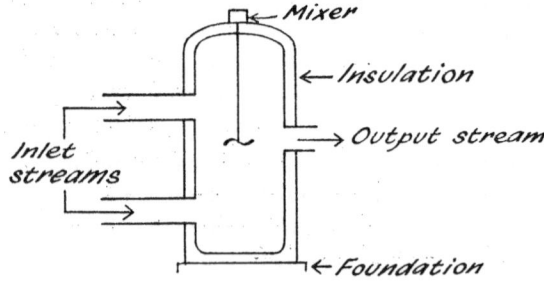

Fig. 5.1

diameter of the cylinder, its volume V taking account of the ends is

$$V = 1{\cdot}7d^3$$

Often there will be upper and lower limits on the size of the reactor and here we will assume that the diameter is constrained to lie between certain values:

$$1.25 \leq d \leq 9{\cdot}67.$$

The cost of the steel for the vessel say C_r depends on the current trade price for steel and the weight W of the vessel. The relationship is estimated as $C_r = 3{\cdot}5W^{0{\cdot}782}$.

We now express the weight W in terms of the diameter d and the temperature of operation. The weight depends on the volume and this is calculated as the product of the surface areas and the thickness. The surface area for the cylinder is $4\cdot70d^2$ square feet and for the ends is $2\cdot46d^2$ square feet. The thickness for the cylindrical walls depends on the required pressure p in the reactor and is given by the equation:

$$t = \frac{p \cdot d}{32,000} + 0\cdot0104 \text{ ft}$$

and for the ends by

$$t = \frac{p \cdot d}{64,000} + 0\cdot0104 \text{ ft.}$$

The reactor must operate at a pressure above the vapour pressure of the reactants which is given by

$$p = 3\cdot3 \times 10^{-6}T^3 \text{ p.s.i.}$$

(T being the reactor temperature). Also, in pressure vessel design, a minimum of 50 p.s.i. and a design pressure 25 p.s.i. greater than the operating pressure are used. Thus

$$p = 50 \text{ p.s.i. if } T \leq 200°F$$
$$= 25 + 3\cdot3(10^{-6}T^3), \text{ for } T > 200°F.$$

Therefore C_r can be expressed in terms of d and T and, since d and V are related, in terms of T and V.

The costs of insulation C_i are dependent on temperature and may be expressed as

$$C_i = 0 \text{ if } T \leq 200°F$$
$$= (17\cdot1 + 1\cdot33 \times 10^{-2}T)d^2 \text{ for } T > 200°F.$$

The cost C_s of a mixer motor depends on the volume it is required to mix and may be expressed as

$$C_s = 255V^{0\cdot3}.$$

Finally, the costs of construction consist of a foundation cost C_m which has a constant component of 1000 and a factor proportional to the diameter d:

$$C_m = 1000 + 100d.$$

Thus the capital costs may be expressed as $C(V, T) = C_r + C_i + C_s + C_m$ and all these quantities can be expressed in terms of the two basic design variables V and T as $V = 1\cdot7d^3$. The original objective function $F(V, T)$ is now a non-linear function of the variables V and T and combined with the constraints we have a non-linear optimization problem. Although the expressions may look cumbersome they present little difficulty when the problem is tackled by a non-linear optimization technique combined with a digital computer.

Often a heat exchanger is added to the type of reactor which has been described and this itself may be a subject for design optimization.

*5.4 Design of machinery

In mechanical and electrical engineering optimization, techniques can be used for design purposes in much the same way as in the chemical industry. However, some new considerations arise compared with the chemical mixing and chemical plant fields. Often a mechanical component can be constructed in one of a number of metals and the characteristics and costs of the metals may differ significantly. The whole design optimization calculation may have to be repeated for a selected set of materials.

A rather different design objective which is encountered in engineering is to produce standard parts which will suit a variety of machines. Here it is necessary to balance the convenience of having a very flexible component, at higher cost per component than necessary for any particular application, against the cost of designing one-off parts constructed more simply for each particular application with the associated delays and increased costs of small-batch production. Furthermore, where there is some degree of standardization useful records can be kept of optimal design characteristics. As each design optimization study takes place on a class of products the results of the studies and the values of the design variables which satisfy various property levels on the product can be recorded. If this information is assembled and tabulated efficiently it can provide a rapid means of assessing the feasible region and even the optimum design characteristics for new products without actually performing a full design study. It is a rationalization and extension of what the designer refers to as his 'intuitive feel' for the design process. The standardization scheme can be applied to small components or even large systems. The next example shows how design optimization techniques can be applied to a range of transformers.

Example 5.3

An air-cooled, two-winding transformer rated 540 VA (1080 VA for both windings together) is to be designed. The manufacturer requires that the full load losses should be no more than 28 watts in order to equal his competitors' claims on efficiency. The transformer should not overheat, i.e. surface heat dissipation should not exceed 0·16 watts/cm². It has been decided that the transformer will be of the shell type illustrated in Fig. 5.2. The transformer is to be designed to minimize initial (or capital) cost and operating cost.

The problem variables are denoted by x_1, x_2, x_3, x_4, x_5, x_6 which, as shown in the diagram stand for the following measurements and quantities:

x_1 = width of core leg in cm

x_2 = width of winding window in cm

x_3 = height of winding window in cm

x_4 = thickness of core in cm

x_5 = magnetic flux density

x_6 = current density.

The objective function consists of the sum of the initial cost plus the operating cost. The initial costs are primarily material costs which depend on the metals used and the volume of the components. If c_1 denotes the price of transformer steel and c_2 denotes the price of copper for the windings, the initial costs $I(X)$ are

$$I(X) = c_1 \text{ (volume of core)} + c_2 \text{ (volume of windings)}$$

$$= c_1(2x_1x_4(x_1+x_2+x_3)) + c_2\left(2x_2x_3\left(x_1 + \frac{\pi}{2}x_2 + x_4\right)\right).$$

The operating costs depend on the economic life of the system T (about 20

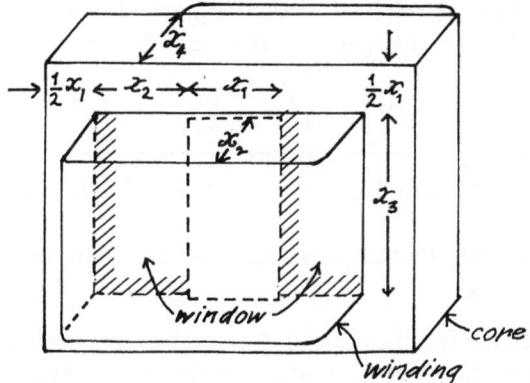

Fig. 5.2

years), the cost of electrical energy c_3, the load factor for the transformer L (about 0·77) and the losses dissipated as heat. This cost $H(X)$ is expressed as

$$H(X) = [\text{losses in core} + \text{(load factor) (losses in winding)}] \ T.c_3$$

$$= [a \cdot \text{(volume of core) } x_5^2 + L \cdot b \text{ (volume of winding) } x_6^2] \ T.c_3 \text{ (where } a$$
and b are technical constants depending on the material)

$$= \left[a(2x_1x_4(x_1+x_2+x_3))x_5^2 + b \cdot L\left(2x_2x_3\left(x_1 + \frac{\pi}{2}x_2 + x_4\right)\right)x_6^2\right] T.c_3.$$

Thus the objective function is expressed as a function of the problem variables:
$$F(X) = I(X) + H(X).$$

The first constraint is that the VA rating must be 1080. The rating is dependent on the product of the area of the core x_1x_4, the area of the winding x_2x_3 and the magnetic flux and current densities.

$$VA = 2ex_1x_2x_3x_4x_5x_6$$

where e is a technical constant incorporating the space factors of iron in the core and conductor material in the winding. The rating condition therefore provides the equality constraint

$$2ex_1x_2x_3x_4x_5x_6 = 1080.$$

The second constraint requires that the full load losses must be no more than 28 watts. As was mentioned in the derivation of the second part of the objective function, the losses are dependent on the expression

a. (volume of core) $x_5^2 + Lb$. (volume of core) x_6^2.

Therefore the limit on maximum losses is expressed as

$$2ax_1x_4(x_1+x_2+x_3)x_5^2+2bLx_2x_3\left(x_1+\frac{\pi}{2}x_2+x_4\right)x_6^2 \leqq 28.$$

The third constraint is that the surface heat dissipated should not exceed 0·16 watts/cm². The surface heat dissipated depends on the ratio of the full load losses and the heat dissipating surface and is expressed as

$$\frac{2ax_1x_4(x_1+x_2+x_3)x_5^2+2bLx_2x_3\left(x_1+\frac{\pi}{2}x_2+x_4\right)x_6^2}{2x_1(2x_1+4x_2+2x_3+3x_4)+4x_2\left(\frac{\pi}{2}x_2+\frac{\pi}{2}x_3+x_4\right)+2x_3x_4} \leqq 0.16.$$

Lastly it is necessary that all x_i values should be non-negative:

$$x_i \geqq 0, \quad i = 1, ..., 6.$$

This completes the specification of the optimization problem in terms of the six variables. Both the objective and constraint functions are non-linear so that a non-linear optimization technique would have to be used to solve the problem.

5.5 Design and research

Design studies are rarely completed without some research investigations and this is inevitably true in the production of a new product or the development of a new machine. In the aeronautical and space industries it becomes almost impossible to separate the two subjects of design and research. Perhaps the greatest difficulty is to assess the relationship between cost and research as the results of research are subject to a great deal of variation. Usually these costs depend on the number of the units which will be made which in turn depends on the success of the research. However, new technologies have to be discussed—the risk of research has to be taken—and some case studies have been recorded where it has been possible to use optimization techniques. We will report here a simplified version of a launch vehicle design study in which the research costs are drawn from a Lockheed report.

Example 5.4

A single-stage rocket-powered launch vehicle is to be constructed and we wish to minimize the total cost of developing, building and launching the vehicle subject to a variety of performance constraints. The vehicle is assumed to have five identical engines of the bell-shaped exhaust-chamber type. We have the following 7 design variables:

x_1 airframe weight in thousands of pounds

x_2 total inert weight in thousands of pounds

x_3 mass fraction

x_4 total thrust in thousands of pounds

x_5 impulse propellant weight in thousands of pounds

x_6 individual engine thrust in thousands of pounds

x_7 length in feet.

The total costs consist of the research, development and production costs for the airframe and the engine expressed in terms of the design variables. The cost functions expressing these components are clearly difficult to measure and require a careful examination of the records of previous designs. The relevant data must be sorted out and regression techniques applied to the information to establish the precise functional expressions. The following functions for the costs are a simplified version of a Lockheed study reported by Rush *et al*. They illustrate how an attempt has been made to sort out the different components in the total cost.

(*i*) Airframe Research and Development Cost Function, F_1.
This cost includes design engineering, special test equipment, etc.

$$F_1 = \frac{5272 \cdot 8 x_1^{1 \cdot 278} x_3^{2 \cdot 424} x_4^{0 \cdot 387}}{x_2^{0 \cdot 196} x_5^{0 \cdot 990}}.$$

(*ii*) Engine Research and Development Cost function, F_2.
This cost includes engineering test hardware, propellant costs, etc.

$$F_2 = -248 \cdot 0 + 160 \cdot 9 (x_6/10^3)^{-0 \cdot 146} + 282 \cdot 9 (x_6/10^3)^{0 \cdot 648}$$

(*iii*) Airframe Production Cost, F_3.
This includes structure tanks, insulation, engine accessories.

$$F_3 = \frac{185214 x_1^{0 \cdot 332} x_5^{0 \cdot 236} x_7^{0 \cdot 108} (n)^{0 \cdot 162}}{x_3^{-1 \cdot 593}}.$$

where $n = 5$, the number of engines.

(*iv*) Engine Production Cost, F_4.
This cost consists of the hardware system of the engine.

$$F_4 = 0 \cdot 208 x_6 10^3 \neq 2 \cdot 509 (x_6/10^3)^{0 \cdot 736} + 0 \cdot 974 (x_6/10^3)^{-0 \cdot 229}$$

The total costs may therefore be expressed as

$$F_1 + F_2 + F_3 + 5F_4$$

as there are five engines each having a production cost although the research for one engine is sufficient for all five and F_2 is therefore only costed once.

It remains to consider the constraints on the x_i values. First there are five engineering relationships between the problem variables which are fixed design commitments. The airframe weight is defined as equal to half the total stage weight, i.e.

$$x_2 = 2x_1.$$

The thrust of the total system is equal to the thrust of all five engines, i.e.

$$x_4 = 5x_6.$$

The mass fraction is defined as the ratio of the initial launch vehicle weight to the vehicle weight subsequent to the burn-up of the propellant, i.e.

$$x_3 = \frac{x_2 + p}{x_2 + x_5 + p}$$

where p is the pay load weight of the instrument unit.

The propellant weight has a structural relationship with the inert weight, so that the following constraint is applied

$$12x_2 \leqq x_5 \leqq 16x_2.$$

Also the length of the vehicle must lie between fixed limits,

$$125 \leqq x_7 \leqq 150.$$

The next set of constraints aim to ensure adequate performance and correct flight characteristics. The first constraint sets bounds on the specific impulse which is the impulse delivered per unit weight of propellant. Thus if t is the burn-up time the specific impulse is $x_4 t / x_5$, and we require

$$240 \leqq \frac{x_4 t}{x_5} \leqq 290.$$

The mass fraction is an important performance parameter since it has a major effect on the velocity attainable. We constrain it to lie between limits as

$$0 \cdot 25 \leqq x_3 \leqq 0 \cdot 30.$$

The thrust to weight ratio is also of crucial importance and it is constrained so that

$$1 \cdot 2 < \frac{x_4}{x_2 + x_5 + p} \leqq 1 \cdot 4.$$

Finally, we may wish to express the final velocity of the vehicle after burn-up in terms of the variables, and, making some simplifying assumptions about the aerodynamics of the flight path, the velocity is expressed as

$$V = \frac{x_4 t \cdot \bar{g}}{x_5} \log \frac{1}{x_3}$$

where $\bar{g}$ is the average gravity over the flight. Thus if we set the desired velocity V this would act as a constraint on the variables.

This completes the specification of the problem and we now have a non-linear objective function and constraints. The actual application was solved by using the created response surface technique.

*5.6 Instrument design: optimization of optical systems

Instruments play a vital role in industrial processes supplying the information on which technological control is based. The chemical process control problems discussed in Chapter 2, vol II assumed that we had pressure gauges, flow rate meters and thermometers to make measurements on the process and check that the process was operating at the determined optimal settings. It may seem curious that instruments themselves should be the subject of an optimization study, but there are certain type of instruments for which not all parameters are determined and an optimizing calculation is needed.

The prime aim of instrument design is accuracy of measurement. However, the scope for accuracy may depend on the value of the measurement being taken and ideally an instrument is required to operate over a range of possible values with equal accuracy. The design problem is to choose the design parameters to maximize the accuracy over a range of possible situations. These ideas will be illustrated in detail by showing how the design of an optical system can be formulated as an optimization problem.

Example 5.5

The objective of any optical instrument is to provide an accurate image of an object located some distance away at a prescribed point in space. We will consider the object to lie in a plane and the image also to be in a plane. The conversion of the object into an image can be interpreted as a series of tracks starting from each point on the object, and passing through the lens system to form the image. The geometrical representation of these tracks is known as ray tracing. Fig. 5.3 shows a system of four lenses spaced along an optical axis with a ray traced from the object plane to the image plane. The position of the ray at any point along the optical axis can be calculated from the technical details of the lens system.

In general, we suppose that the optical system consists of a series of N lens surfaces each of which will be presumed to be spherical. The properties of the system are defined by the spacing of the surfaces along the axis, their radii of curvature, and the relative refractive indices of the media. Knowing these details, the track or ray can be determined as it is 'bent' or refracted at each lens surface. We wish to choose the system of surfaces so as to minimize the aberration of the image formed of the object. Therefore the design variables of the system consist of three characteristics of each surface:

t_i = distance along optical axis from surface i to surface $(i+1)$

r_i = radius of curvature of surface i

l_i = relative refractive index on either side of ith surface (this is a property of the glass used).

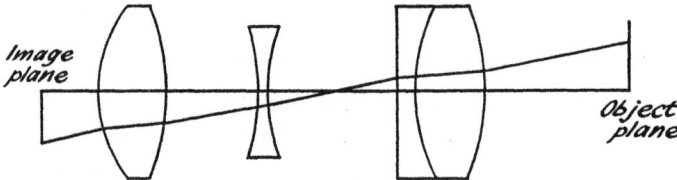

Fig. 5.3

The aberration is measured by the ray tracing procedure. At each surface the position of the ray is defined in terms of its coordinates on the surface and its direction on leaving the surface. Taking the optical axis as the origin for any surface, the coordinates of the ray at surface i can be expressed as (u_i, v_i) where these are the distances in two perpendicular directions from the origin. (u_0, v_0) will be the coordinates in the object plane of a given point on the object from which the ray emerges (Fig. 5.4).

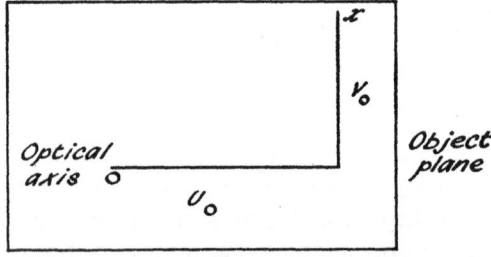

Fig. 5.4

Also the directions of the ray at surface i are the angular directions a_i and b_i between the ray and the two planes through the optical axis perpendicular to one another and to the object plane. The ray is therefore specified at the ith surface by the numbers

(u_i, v_i, a_i, b_i).

The lens system is specified by the parameters t_i, r_i and l_i and these quantities can be used to calculate the components $(u_{i+1}, v_{i+1}, a_{i+1}, b_{i+1})$ by an iterative calculation which we summarize by the symbol h in the relations

$u_{i+1} = h_u(u_i, v_i, a_i, b_i, t_i, r_i, l_i)$

$v_{i+1} = h_v(u_i, v_i, a_i, b_i, t_i, r_i, l_i)$

$a_{i+1} = h_a(u_i, v_i, a_i, b_i, t_i, r_i, l_i)$

$b_{i+1} = h_b(u_i, v_i, a_i, b_i, t_i, r_i, l_i).$

By repeating the whole procedure at each surface the coordinates of the ray in the image plane say (u_N, v_N) will be determined exclusively as functions of (u_0, v_0, a_0, b_0) and all the t_i, r_i, l_i values for $i = 1$ to N. Denoting the N values of t_i, r_i, l_i by the vectors T, R and L, the relations may be written as

$u_N = H_u(u_0, v_0, a_0, b_0, T, R, L)$

$v_N = H_v(u_0, v_0, a_0, b_0, T, R, L).$

If u_0 and v_0 are held fixed these relations for varying angular directions a_0 and b_0 represent a bundle of rays emerging from a particular object point and the requirement for sharp imagery is that this ray bundle should again coalesce at a point in the image space. In other words u_N and v_N ought to depend only upon u_0 and v_0 and not upon a_0 and b_0. Furthermore, in order to secure freedom from distortion so that geometrical similarity is preserved the following relations must hold:

$u_N = Mu_0$

$v_N = Mv_0$

where M is the constant of magnification. These last two equations effectively provide an aberration function which we may measure as the sum of the squared deviations

$f(T, R, L) = (u_N - Mu_0)^2 + (v_N - Mv_0)^2$

and we wish to choose the T, R and L values to minimize this quantity.

It is usually possible to find a large number of optical systems which will minimize $f(T, R, L)$ for the particular (u_0, v_0). However, we would want to avoid all aberrations for a range of possible coordinates in the object plane say $(u_0^{(i)}, v_0^{(i)})$, for $i = 1$ to M, and if the aberration function for point i is $f^{(i)}(T, R, L)$ we want to determine T, R, L to minimize

$$F(T, R, L) = \sum_{i=1}^{M} f^{(i)}(T, R, L).$$

The extent to which we can approach zero in the value of $F(T, R, L)$ determines the quality of the optical system.

In practice there will also be a number of constraints on the t_i, r_i, l_i quantities. Normally these will be inequalities of the form

$G(T, R, L) > 0.$

The constraints must take account of the physical conditions such as the need for the distance between the lenses to be positive, the requirement that all lenses must have sufficient edge and centre thickness and the availability of refractive indices in real glasses.

5.7 Applications in electronics

The electronics industry is a natural source of design optimization problems where the data and behaviour of the systems are reasonably well understood and mathematically defined. Typical applications are the design of transistors, filters and switching circuits. However, there are other areas in which optimization techniques can be used and we will illustrate their application to the important field of non-linear network analysis.

The problem in network analysis is to solve a system of equations to determine currents, voltages, etc., in the separate components. However, rather than tackle the analysis directly as an equation solving problem, the task is sometimes better achieved by a minimization technique. Suppose a network is defined in terms of a set of variables $x_1, x_2, ..., x_N$ which must satisfy a set of simultaneous algebraic equations

$$g_k(x_1, x_2, ..., x_N) = 0, \; k = 1, 2, ..., M.$$

The equations will describe the voltage-current relationships in a non-linear resistive network. It may be very difficult to solve the equations directly and instead we can minimize the function F consisting of the sum of the squares of the g_k functions:

$$F(x_1, x_2, ..., x_N) = \sum_{k=1}^{M} [g_k(x_1, x_2, ..., x_N)]^2.$$

The effect of squaring ensures that when F is minimized we cannot determine values for which g_k is negative. Then the values of the problem variables for which $F = 0$ solve the equations.

Example 5.6

A simple resistor diode network is illustrated in Fig. 5.5 and the problem is to find the diode current flowing. Let i_u denote the unknown diode current. The descriptive non-linear algebraic equation for the network is:

$$v = v_u + i_u \cdot R$$

where v is the source voltage, v_u is the voltage drop across the diode and R is the resistance.

The diode current i_u and the diode voltage v_u are known to be related by the equation

$$i_u = 10^{-9}(e^{40v_u} - 1).$$

The original equation can be rewritten using this connection between i_u and v_u as

$$\frac{v - v_u}{R} - 10^{-9}(e^{40v_u} - 1) = 0.$$

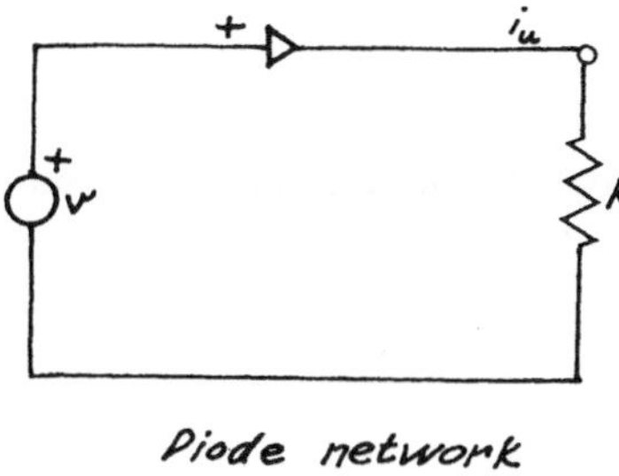

Piode network

Fig. 5.5

Knowing v and R, we can now square the quantity on the left-hand side of the equation and use the method of steepest descent to determine the minimum. This will give the value of v_u and hence the value of i_u can be determined by the characteristic equation for the diode.

*5.8 Layout problems: the network crossings problem

A rather different kind of design problem from the types which have been considered so far occurs in planning the layout of a number of interconnected items. These layout problems arise in a variety of contexts. In a factory it is desirable to locate the machine facilities so as to minimize the internal transportation of work and to improve the flows of jobs. The design of a large chemical complex will require numerous connecting pipes between the various equipments and it is important to organize a good pipe layout. The pipe layout would need to consider the necessary equipment separation, the ease of access to instrumentation and the additional resistance encountered if too many bends occur or extra piping was used. Perhaps the most developed area for layout design is in the positioning of interconnected logic elements on a circuit board to achieve defined wiring objectives. One of the objectives in this field is to place the logic elements in a set of possible positions on the board so as to minimize total wire length: this is known as the backboard wiring problem, Another, more difficult problem, is to arrange the layout so as to minimize the number of crossings between the connecting wires. This problem will be examined in detail to illustrate a special formulation which may be appropriate to layout problems.

Example 5.7

The network of an electronic device consists of a number of logic elements with specified connections between the elements. We will consider the simplest situation in which each element is a single pin and the connections are specified as a list of connections between pairs of elements. Suppose these were 5 elements numbered 1, 2, 3, 4, 5 with eight connections between them as listed below

1—2	2—3	3—4
1—3	2—4	4—5
1—5	2—5	

Fig. 5.6 shows how the network could be layed out on the board. With the

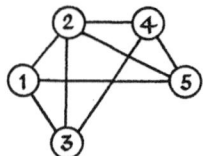

Fig. 5.6

elements in these positions, and the connections drawn as shown three crossings occur between the connections: between (1—5) and (2—3), between (1—5) and (3—4), between (2—5) and (3—4). The layout problem is to place the elements on a rectangular board and decide the routes for the connections on the board so as to minimize the number of intersections or crossings between the connections. Where a crossing takes place it will not be possible to create or 'print' the connection automatically by depositing a thin line of conductor on the board as an electrical contact will occur. Wherever such crossings do occur on the final design it will be necessary to introduce a

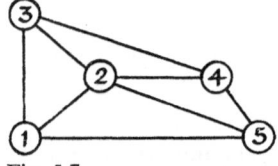

Fig. 5.7

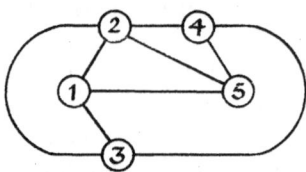

Fig. 5.8

special piece of insulated wire or a layer of insulation, which is inconvenient and comparatively expensive. In the case of the network of Fig. 5.6 we can eliminate the crossings by repositioning element number 3 or re-routing its connections as shown in Figs 5.7 and 5.8.

Although it is clear visually how to redraw the network of Fig. 5.6 so as to eliminate the crossings it is not immediately clear how to perform this operation computationally. We need a numerical representation for the network which enables us to calculate the number of crossings by a formula and facilitates the manipulation of the network to reduce crossings. This can be achieved by a permutation representation of the problem after the network has been drawn in a special way. The network is redrawn so that all the elements lie on a line in the plane (which we will refer to as the element line) and all the connections are drawn as semicircles connecting pairs of elements, the semicircles lying either above or below the element line. If the network of Fig. 5.7 is redrawn in this way it will take the form illustrated in the diagram of Fig. 5.9 in which the numbers marked on the connections correspond to the list of original connection pairs in their order in the list. It can be shown that this deformation of the network can nearly always be achieved so that the optimum of the resulting network in the new representation is equivalent to the optimum in the original form.

In this new form it is comparatively simple to represent a network as a permutation problem. If there are N elements the list of elements along the element line working from left to right can be represented by the permutation

$$[P] = [p_1, p_2, ..., p_N]$$

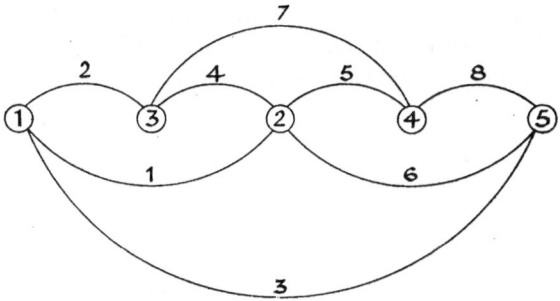

Fig. 5.9

where p_j is the element in the jth position. Also we can record whether the connections lie above or below the element line by the vector

$$(Q) = (q_1, q_2, ..., q_M)$$

where $q_i = +1$ if the ith connection lies above the element line and $q_i = -1$ if the connection lies below the element line. The complete layout of the network is thus expressed by the combined information

$$[P; Q] = [p_1, p_2, ..., p_N; q_1, q_2, ..., q_M].$$

For the network of Fig. 16.9 the representation is

$$[P; Q] = [1, 3, 2, 4, 5; -1, +1, -1, +1, +1, -1, +1, +1].$$

The first set of numbers record the element sequence and the second set state the connection pattern. For instance as connection 1 is below the line $q_1 = -1$ and as connection 2 is above $q_2 = +1$, etc.

We now need a formula to measure the number of crossings in the network in this form. Associated with any permutation $[P; Q]$ two ancillary matrices $A(i, j)$ and $B(i, j)$ can be calculated where $A(i, j) = 1$ if there is a connection from the element in position i to the element in position j above the element line, and $A(i, j) = 0$ otherwise, and $B(i, j)$ is defined similarly for connections

E

below the line. Then the number of crossings associated with the permutation $[P; Q]$ can be expressed as

$$F[P; Q] = \tfrac{1}{2} \sum_{i=1}^{N-1} \sum_{j=i+1}^{N} A(i, j) \left\{ \sum_{k=1}^{i-1} \sum_{l=i+1}^{j-1} A(k, l) + \sum_{k=i+1}^{j-1} \sum_{l=j+1}^{N} A(k, l) \right\}$$

$$+ \tfrac{1}{2} \sum_{i=1}^{N-1} \sum_{j=i+1}^{N} B(i, j) \left\{ \sum_{k=1}^{i-1} \sum_{l=i+1}^{j-1} B(k, l) + \sum_{k=i+1}^{j-1} \sum_{l=j+1}^{N} B(k, l) \right\}.$$

This formula is devised by considering any crossings of the form shown in Fig. 5.10. In the diagram the connection ($p(i)$ to $p(j)$) intersects the connection ($p(k)$ to $p(l)$) and this is contained in the formula as the product

$$A(i, j) \cdot A(k, l).$$

Only if both $A(i, j)$ and $A(k, l)$ were unity would we obtain a unit contribution to the total number of crossings. Although the formula may appear complex, it is quick to evaluate on a computer because it consists entirely of additions.

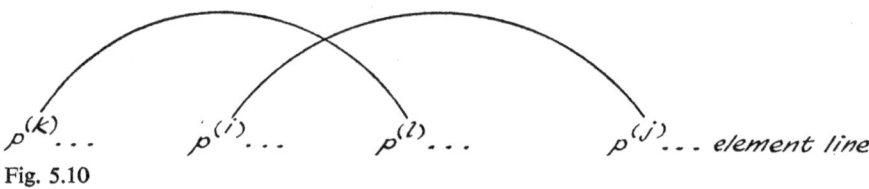

Fig. 5.10

We now have to choose a set of exchanges to define the local optimum. Let us define the set of exchanges as all single shifts of any element, given that, in the new position, the routes for the connections from the altered element are redrawn optimally. Therefore given the permutation $[P; Q]$, if we move the element in position i to position j adjusting the positions of the other elements as for a single shift exchange and redraw the connections from the element which was in position i in an optimal way to minimize crossings, we can denote the resulting permutation by $[P(i, j); Q']$. The permutation $[P; Q]$ is a locally optimal permutation with respect to such exchanges if $F[P; Q] < F[P(i, j); Q']$ for all i and j.

A locally optimal permutation can be constructed by the usual two stage procedure. An initial permutation is determined by introducing the elements into the permutation one by one. The first element to be chosen is the element with the most connections and it is placed in the first position. Subsequently, the next element to be selected is the element which has most connections to the elements already placed in the permutation. The selected element is tried in each position along the node line and it is placed in the position for which the increase in the number of crossings is least, with its connections drawn as semicircles above or below the line according to whichever route offers fewer crossings. When the initial permutation is complete it is then exchanged into a locally optimal permutation by applying the exchange sequence until no further improvements can be made.

To illustrate the procedure, the sequence of iterations for constructing the network is reproduced graphically on a small example. The example network consists of 7 elements and the connections between the pairs of elements is as follows:

1—2	2—4	3—4	4—6	5—7	6—7
1—3	2—5	3—5	4—7		
1—4	2—7				
1—5					

Stage 1

The initial sequence of elements is selected and positioned as shown in the following diagrams:

Selection 1: element 4 into position 1.

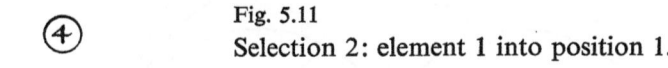

Fig. 5.11

Selection 2: element 1 into position 1.

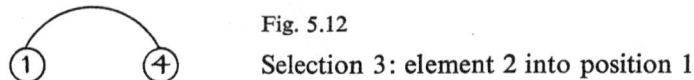

Fig. 5.12

Selection 3: element 2 into position 1.

Fig. 5.13

Selection 4: element 3 into position 1.

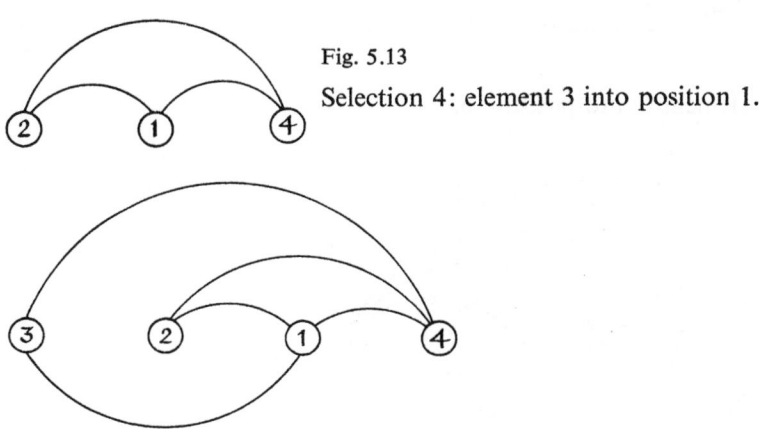

Fig. 5.14

Selection 5: element 5 into position 2.

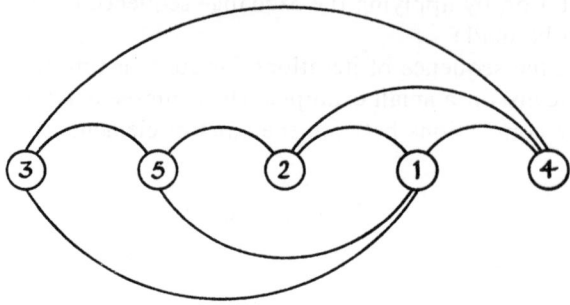

Fig. 5.15

Selection 6: element 7 into position 3.

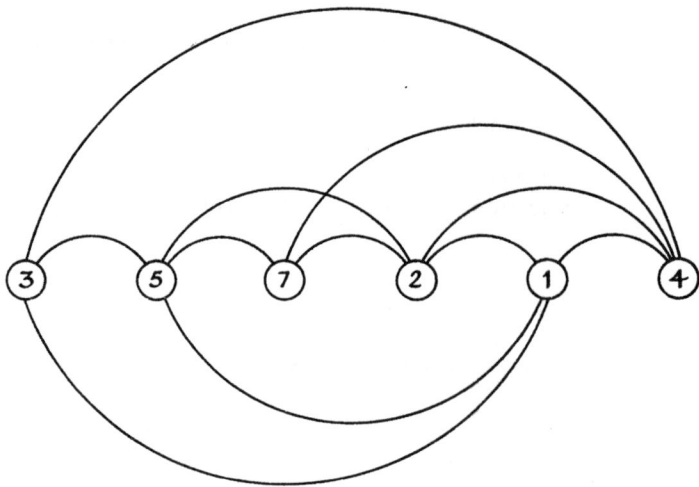

Fig. 5.16

Selection 7: element 6 into position 5.

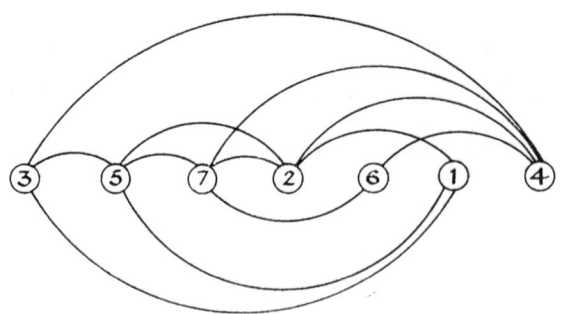

Fig. 5.17

Stage 2

The final sequence of elements from stage 1 is now adjusted by selecting element 2 and repositioning it in position 5. This gives a network with no crossings and so completes the process.

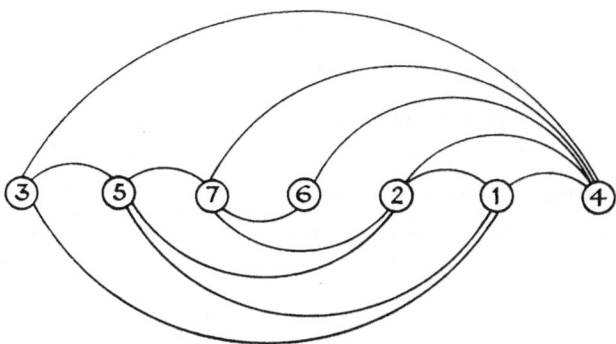

Fig. 5.18

The method was also applied to some networks for which the minimum number of crossings is known. First it was applied to some complete graphs which are networks in which all elements are connected to all other elements. The results are given in Table 5.1 showing that the optimum results are obtained by the permutation procedure in all except one case.

Table 5.1

Number of crossings in complete graphs

Number of elements	Number of crossings		
	Initial permutation	Final permutation	minimum
5	1	1	1
6	3	3	3
7	9	9	9
8	20	18	18
9	36	36	36
10	62	60	60
11	102	100	100
12	157	150	150
13	231	227	225
14	325	315	315

The method was also tested on 18 randomly generated networks. The number of elements in these networks was set at 10, 15 and 20 and the number of connections was varied so that the average number of connections per element ranged over the values 3, 6 and 9. A Monte Carlo method was also tried on the networks in which the number of crossings in the best network after 10, 50 and 250 trials was recorded. The results are shown in Table 5.2.

Table 5.2

Case	Number of elements	No. of connections		Number of crossovers				
				Permutation procedure		Monte Carlo trials		
		Average	Total	Initial	Final	10	50	250
1	10	3	15	0	0	2	1	0
2	10	3	15	1	1	3	2	1
3	10	3	15	0	0	2	1	0
4	10	6	30	18	12	22	22	17
5	10	6	30	16	9	19	19	17
6	10	6	30	16	14	23	18	16
7	15	3	23	2	0	7	7	6
8	15	3	23	0	0	13	6	4
9	15	3	23	1	0	15	7	4
10	15	6	45	38	26	76	63	50
11	15	6	45	31	26	73	61	54
12	15	6	45	41	19	66	58	52
13	15	9	68	126	101	185	173	159
14	15	9	68	121	105	177	167	158
15	15	9	68	132	101	191	171	156
16	20	3	30	10	3	22	22	19
17	20	6	60	59	57	163	146	120
18	20	9	90	233	183	370	368	334

The Monte Carlo trial networks were constructed partly randomly and partly systematically. For a network of N elements, a random permutation of the integers 1 to N was drawn for each trial. Having obtained the element positions as represented in the random permutation all the connections were drawn as single semicircles above or below the line whichever was more advantageous taking each element in turn. This meant that the randomness was transmitted through the element positions, but that the connections were drawn systematically in an efficient manner.

Table 5.2 shows how very considerable gains can be made by placing the elements systematically along the line rather than randomly, despite the efficient routing of the connections which was built into the Monte Carlo method. Once the locally optimal permutation had been obtained the network would be deformed so that the elements lay over a rectangular region to correspond to a circuit board shape. This can be achieved without increasing the number of crossings by reversing the deformation scheme which was used to convert a network into the special form for permutation representation.

REFERENCES

Aris, R. 1961. *The Optimal Design of Chemical Reactors*. Academic Press, New York.
Buckler, E. J. and Kristensen, I. M. 1967. The use of response equations for appraisal of elastomers. *J. Inst. Rubber Ind.*
Davies, O. L. 1954. *The Design and Analysis of Industrial Experiments*. Hafner Publishing Co., New York.
Feder, D. P. 1966. Lens Design viewed as an Optimization Process. In *Recent Advances in Optimization Techniques*. Eds. A. Lavi and T. P. Vogl. Wiley
Garside, R. G. and Nicholson, T. A. J. 1968. The backboard wiring problem. *Proc. I.E.E.*
Gelbin, D. 1964. Optimizing the design of a multi-stage ammonia reactor. *Chem. Eng. Progr. Symp. Series No. 60.* Am. Inst. Chem. Eng.
Godwin, G. L. 1954. Optimum machine design by digital computer. *A.I.E.E. Trans.*
Nicholson, T. A. J. 1968. A permutation procedure for minimizing the number of crossings in a network. *Proc. I.E.E.*
Rush, B. C., Bracken, J. and McCormick, G. P. 1967. A non-linear programming model for launch vehicle design and costing. *Opns Res.*, **15**, 185.
Schinzinger, R. 1966. Optimization of Electromagnetic System Design. In *Recent Advances in Optimization Techniques*. Eds. A. Lavi and T. P. Vogl. Wiley
Singer, E. 1962. Simulation and optimization of oil refinery design. *Chem. Eng. Progr. Symp. Series No. 37.* Am. Inst. Chem. Eng.
Temes, G. C. and Calahan, D. A. 1967. Computer-aided network optimization—the state of the art. *Proc. I.E.E.*, **55**, 1832.

Exercises on Chapter 5

1 An alloy is made by mixing three types of metal ore, *A*, *B* and *C*. The manufacturer is interested in two properties of the alloy: a ductility factor (*D*) and a corrosion factor (*E*). These properties are measured for two proportions of the two metals *A* and *B* with the following results on a unit mix. (It is presumed that the proportion of the metal *C* has no effect on these properties.)

Table 5.3

Proportion of metal *A*	Proportion of metal *B*	Level of property *D*	Level of property *E*
$\frac{1}{4}$	$\frac{3}{4}$	3	4
$\frac{3}{4}$	$\frac{1}{4}$	6	3

The manufacturer uses these results to express the values of the properties in the form $a_1 x_1 + a_2 x_2$ where x_1 and x_2 are the proportions of the metals *A* and *B* in a unit mix. The costs of the three metals *A*, *B* and *C* are in the ratio 2 : 3 : 1. A customer requires an alloy with a ductility of at least 5 units and a corrosion factor of at most 4 units. Determine the best proportions in which the manufacturer should mix the alloy.

2 A chemical engineer is designing a reactor and he has decided that it will consist of a central cylinder and a hemispherical shell at either end. The production requirements are such that the volume of the reactor must be between 500 and 1000 cu.ft. The length of the cylinder must be between 6 and 12 ft. The estimated profit on a cubic foot of product is £0·25 and the holding time of the material is 2 hours. The reactor is expected to last 8 years. The costs of the reactor consist of the material costs, insulation costs, mixer motor costs and foundation costs. The material and insulation costs depend on their volume requirements which are approximated as the product of the surface area and the thickness of the material. The thickness is fixed at 3 inches for the steel shell and 9 inches for the insulation. Steel costs £10 cu.ft and insulation £5 per cu.ft. The cost of the mixer motor is proportional to the square root of the volume, a motor for a 100 cu.ft reactor costing £500. The cost of the foundation is a basic £500 plus £100 for every foot of radius.

Formulate in detail the problem of designing the shape of the reactor so as to maximize total profit.

3 A manufacturer is designing an instrument which can serve a number of purposes. The instrument is defined by the scale $(0, t)$ in which it will operate. The lower limit 0 is fixed but the upper limit t is a variable. The manufacturer has to meet a demand for the instrument which varies with respect to the desired upper limit. If an application requires an instrument with a range $(0, s)$, any instrument with an upper limit $t \geqq s$ will serve the purpose. The manufacturer wants to design a standard stocked instrument with range $(0, t)$ which will cost him $(c_0 + c_1 t)$ to serve most of the demand, and make special purpose instruments at extra set-up cost and manufacturing cost (proportional to the range required) for any demand for instruments outside the standard. The cost of these special instruments is $(c_0' + c_1 s)$ where $c_0' > c_0$. The demand for the upper limits for the instrument falls steadily over the interval $(0, T)$, there being no demand for an upper limit beyond T. Show how the manufacturer could determine the limit for his standard design in order to minimize total costs.

*4 N machines are to be located on a factory floor. N possible positions have been designated for the machines and the distance between position i and position j is $d(i, j)$. The planned process sequences have been assessed and the typical work flows which will go from machine i to machine j are given by relative flow coefficients $v(i, j)$. The management wish to locate the machines to minimize the average total distance over which work will have to be transported between machines. Examine how this problem can be formulated as a permutation problem and suggest what kind of exchanges might be suitable for defining the local optimum.

*5 A circuit layout is to be designed so as to minimize the number of crossings between connecting wires. The circuit has a special feature in that all the elements must lie along one edge of the board. This means that instead of being able to draw the connections above or below the element line as was possible in Example 5.7, the connections must be drawn as semi-circles on one side of the element line. Represent the network as a permutation and, using the first set of data in Example 5.7 with 5 elements and 8 connections, determine a network for which the crossings are minimal with respect to adjacent exchanges.

6 Plant renewal and location

6.1 Long-term planning problems

In some of the previous chapters we have classified the applications by the planning periods to which they relate. In Chapter 4, Vol. II we designated stock control and production scheduling to be medium-term planning problems as the decisions related to the setting of production patterns over periods measured in months. Previously we had studied really short-term problems—machine sequencing and process control—where decisions have to be made usually on a daily basis. In this chapter we embark on the really long-term problems of deciding the size of a plant and where to place it. These are capital investment decisions made in periods of expansion perhaps once a year or even every five years. For the short- and medium-term problems a dominant constraint has been the given capacity available, but here we wish to determine the size of the capacity which should be created. Although the plant size and location decisions occur infrequently, the economies to be made by the correct judgment of capacity and siting are enormous. As a general rule, the longer the period to which the decision relates, the more there is at stake.

The simplest context for the replacement decision is the choice of when to replace a single machine, in which demand and technological change are not considered. The running costs will tend to rise as the machine ages and the salvage value will fall so that there comes a point in time at which it is worth while to replace the machine despite the capital costs which are incurred in purchasing a new machine. There are a variety of criteria which may be used

124

to make this decision and they are discussed in the next section. The machine replacement problem becomes more realistic and more difficult when predictions about technological improvement are included. It is then not nearly so clear when a renewal is due as the costs of future machines will be falling in an unsteady manner. Fortunately dynamic programming can be used to handle this problem in an elegant way and this will be described in a subsequent section.

The concept of replacement in the strict sense implies that we wish simply to maintain the function of an existing capital asset such as a machine. More generally we may wish to increase the physical capacity as an expansionary investment. In this case the market demand will be incorporated in the model and plant renewal decisions will be dealing with the total composition of the firm's manufacturing facilities. The national electricity generating system provides an interesting and important area for this kind of investigation: the future demand is reasonably well known, technological improvements and the choice of plant types are predicted and the system operates as an integrated unit. We will study the 'industry' renewal and investment problem as related to the electricity generating system in Section 6.4.

Wherever an expansionary investment is being implemented there may be a choice of location for the new plants. Here it will be necessary to consider both the increase in demand and its distribution at different geographical points. Often it may be best to locate a single plant at its best point and allow it to produce for the whole market, thus invoking the maximum economies of scale for the production unit. On the other hand, if the market is widely scattered it may be better to locate two plants and thus reduce the total transportation costs. The general problem of deciding plant location, the production scale of each plant and the distribution policy are discussed in Section 6.5

6.2 Criteria for replacing a single machine

The simplest version of the replacement problem considers a static market and no technological change. Although these assumptions are unrealistic, the elementary model for this situation enables the principal issues in the subject to be identified. We will consider the case of a single machine and study the question of when it should be replaced by another machine for which the cost characteristics are known. As there are no market considerations the size of the machine can be ignored.

There are three main cost variables associated with a machine which alter as the machine ages. First, the revenue from the use of the machine will fall as time goes on in the manner illustrated in Fig. 6.1. Starting at some high value it will decrease as the equipment deteriorates. The revenue forms a productivity measure. The revenue from the machine in year t will be denoted by $r(t)$. Secondly, the operating cost will increase as the machine gets older and this may take account of increasing maintenance charges associated with more frequent failure. The operating cost will increase over time as illustrated in Fig. 6.2. It is denoted by $u(t)$. As the current revenue is falling and the

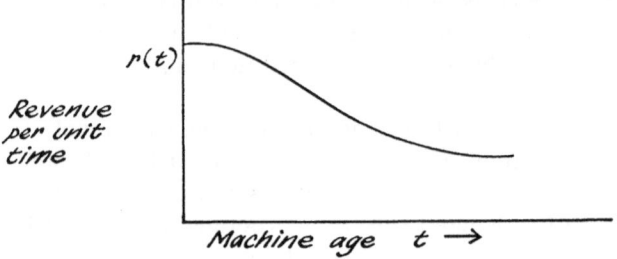

Fig. 6.1

current costs are increasing, the net return $r(t) - u(t)$ from the machine will also be declining.

On the capital side, the capital costs of the existing machine are 'sunk' and cannot be considered in the replacement decision. However, the salvage value

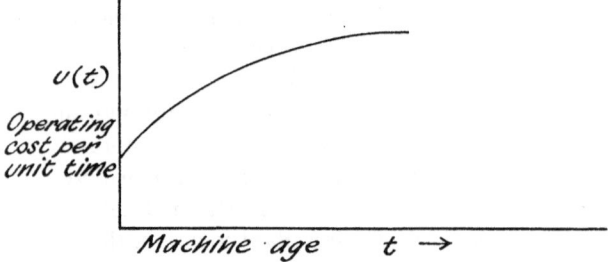

Fig. 6.2

of the machine will be steadily falling as the machine ages. Denoting the salvage value by $s(t)$, a typical salvage value curve is illustrated in Fig. 6.3.

The problem facing a company is to decide when a given machine purchased at a given time should be replaced. This depends on the alternative machines which are available and their costs, and the cost measure which the company

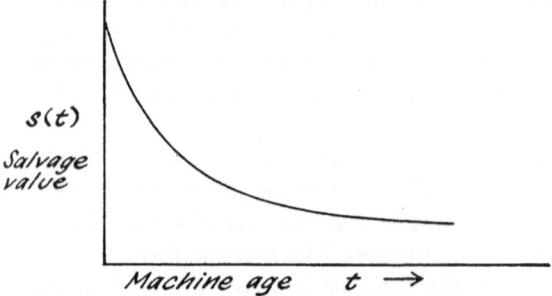

Fig. 6.3

adopts. Suppose that a company owns a machine in year $t = 0$ which has revenue, operating cost and salvage value $r(t)$, $u(t)$, $s(t)$ in year t. The company wishes to maximize returns over the period $t = 0, 1, 2, ..., T$. Let the cost of a new machine be A, its salvage value be S after a specified lifetime L, and let the total revenue less operating cost over this period be R. Then the average annual return from the new machine is

$$a = \frac{R-A+S}{L}.$$

If x denotes the year in which the machine is replaced, the total returns over the period $0 \leq t \leq T$, are

$$F(x) = \sum_{t=1}^{x} (r(t) - u(t)) + s(x) + (T-x)a.$$

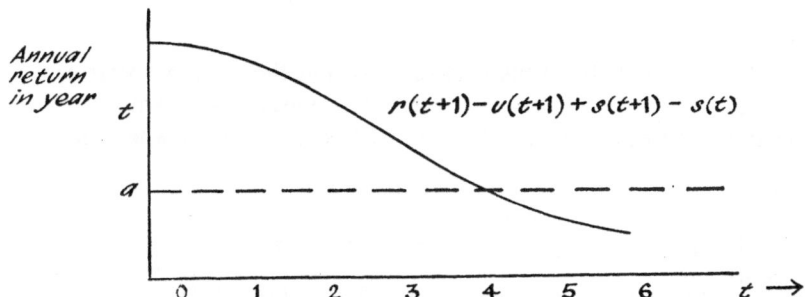

Fig. 6.4

We wish to choose x to maximize $F(x)$. Taking the discrete derivative:

$$F(x+1) - F(x) = \sum_{t=1}^{x+1} (r(t) - u(t)) + s(x+1) + a(T-x-1)$$

$$- \left\{ \sum_{t=1}^{x} (r(t) - u(t)) + s(x) + a(T-x) \right\}$$

$$= r(x+1) - u(x+1) + (s(x+1) - s(x)) - a.$$

This is positive so long as

$$r(x+1) - u(x+1) + s(x+1) - s(x) > a.$$

Therefore it is optimal to replace the plant in the year when the net revenue from the next year less the incremental charge in salvage value next year is less than the average annual return from the new machine.

Fig. 6.4 illustrates the replacement decision showing that replacement will occur in year 4.

Often it is difficult to measure returns explicitly and the replacement decision may be based entirely on the cost structure. This is specially true where no consideration is being made of the market. The criterion which has just been derived can then be rewritten. We will choose the replacement year as the first year in which the annual costs, including the decrease in salvage value, exceed the average cost of the new machine.

It may be very difficult to assess the precise cost characteristics of alternative machines, and a separate question may be investigated of when to replace the existing machine with an identical machine. Suppose the machine has a capital cost of C and it is replaced after x years with an identical machine; then, using the above notation, the average costs are the net capital cost plus the operating costs:

$$F(x) = \frac{1}{x}\left[C - s(x) + \sum_{t=1}^{x} u(t)\right].$$

This function can be minimized directly by evaluation, as illustrated in Example 6.1.

Example 6.1

A machine tool owner finds from his past records that the costs per year of running a machine whose purchase price is £2000 are given as shown in Table 6.1 together with the salvage value. He wishes to decide at what age

Table 6.1

Age in years	1	2	3	4	5	6	7	8
Operating costs	333	400	466	600	766	933	1133	1333
Salvage value	1000	500	250	125	66	66	66	66

the machine should be replaced with the criterion of minimizing average annual cost.

We can now compute a table of costs showing what will be the average costs per year under the different replacement policies. These are given in Table 6.2, where the total capital costs refer to the purchase cost £2000 less the salvage value. From the table it is clear that the average costs per year are

Table 6.2

Replace at end of year	Total running costs	Total capital costs	Total costs	Average cost per year
1	333	1000	1333	1333
2	733	1500	2233	1116
3	1200	1750	2950	983
4	1800	1875	3675	919
5	2566	1934	4500	900
6	3500	1934	5434	906

at a minimum if replacements are made every fifth year. This minimization has been achieved by a straightforward evaluation from the calculated table.

So far we have only considered two possible cost criteria: the maximization of returns over a fixed period of time and the minimization of average annual costs. But these are not the only criteria. Perhaps the best criteria, and certainly the one most commonly used, is to determine the replacement policy so as to minimize the present value of the discounted future costs or maximize the present value of future returns.

The discounting procedure to obtain present values is a very common practice in financial calculations. The present value of an amount of money a in one year's time is $\dfrac{a}{1+r}$ if r is the rate of interest; this follows from the fact that if an amount $\dfrac{a}{1+r}$ was invested for a year, the total value at the end of a year would be

$$\frac{a}{1+r}(1+r) = a.$$

Similarly the present value of an amount a in two year's time is $a/(1+r)^2$.

For brevity we denote the factor $\dfrac{1}{1+r}$ by d and refer to this as the discount factor. If a series of payments $u(t)$ are made in years $t = 0, 1, 2, \ldots$, and the payments can be considered to take place at the beginning of each year, the present value $P(t)$ of the total expenditure over t years is:

$$P(t) = u(1) + du(2) + d^2u(3) + d^3u(4) + \ldots + d^{t-1}u(t).$$

This method of discounting costs to obtain a present value of total costs is equally well applied to discounting returns to obtain a present value of total returns.

An advantage of the discounting feature is that the horizon for the replacement study can be taken as infinity as the sum of the costs will always converge to a finite number. Suppose a manufacturer purchases a machine in year 0. In year t the estimated return is $r(t)$, the operating cost is $u(t)$ and the salvage value is $s(t)$. C is the cost of buying a new machine. The manufacturer wishes to determine how frequently the machine should be replaced.

Assuming that there is no technological change, the optimal policy will be to replace the machine in a regular cycle say every T years. The value of T is the problem variable. The discounted returns in each of these T years will be.

year 1 $r(1) - u(1)$

year 2 $(r(2) - u(2))d$

year 3 $(r(3) - u(3))d^2$

$\cdots$

$\cdots$

year T $(s(T) - C + r(1) - u(1))d^{T-1}$.

Addition of these quantities gives the total discounted returns $F(T)$ for the first machine:

$$F(T) = \sum_{i=1}^{T-1} (r(i) - u(i))d^{i-1} + (s(T) - C + r(1) - u(1))d^{T-1}.$$

Therefore the total returns for the whole chain of replacements occurring at times 0, T, $2T$, $3T$, ..., is

$$F(T) + F(T) . d^T + F(T) . d^{2T} + F(T) . d^{3T} + \dots$$

$$= \frac{F(T)}{1 - d^T}$$

by the geometric summation formula. Thus the objective function for the total discounted returns of replacements occurring at intervals of T is

$$F(T) = \frac{\sum_{i=1}^{T} (r(i) - u(i))d^{i-1} + (s(T) - C + r(1) - u(1))d^{T-1}}{1 - d^T}$$

The optimal machine life is obtained by finding the value of T for which $F(T)$ is a minimum. Although in certain circumstances it might be possible to use an optimization technique to minimize $F(T)$, it would usually be satisfactory to evaluate $F(T)$ for a chosen range of values of T and obtain the optimum by inspection. If the formula was being used to compare two machines, we would evaluate the optimal $F(T)$ values in both cases and compare the results.

The optimal value of T which is obtained by the discounting model may differ significantly from the value which would be determined if the criterion was to minimize average annual costs. A further practical factor which will influence the optimal replacement period is the tax allowances on capital investment. The effects of these factors and the different criteria are illustrated in the following case study applied to fork-lift trucks, reported by Eilon. *et al.*

Example 6.2

A company owning a fleet of fork-lift trucks wishes to determine the optimum working life of an average truck in the fleet under a variety of financial criteria.

The price of a fork lift truck is £1700 (except for batteries which are considered separately). The re-sale values at various ages have been estimated for a life up to ten years as shown in the Table 6.3. An experimental curve of the form Ke^{-at} has been fitted to this data. The parameters K and a were estimated by regression to be 1300 and 0·4, so that the salvage value $s(t)$ in year t is estimated as

$$s(t) = 1300e^{-0.4t}.$$

The values of $s(t)$ are also shown in Table 6.3.

The maintenance costs were obtained by taking a sample of ten trucks and fitting a regression line of the form

$$u(t) = a+bt$$

to the data. The value of a was 49·9 and of b was 24·9 so that the operating or maintenance costs were represented in year t as

$$u(t) = 49·9+24·9t.$$

It is assumed that there is a standard corporation tax of 40 per cent and that there is a system of allowances for capital expenditure. In the year of purchase there is an investment allowance of 30 per cent. This is accompanied by an initial allowance of 10 per cent plus an annual allowance of 25 per cent in the first year. The 'written down value' of the investment subtracts the annual allowances from the capital cost. Therefore if the capital cost is C, the written down value at the end of the first year is

$$C-0·10C-0·25C = 0·65C.$$

Table 6.3

Year	1	2	3	4	5	6	7	8	9	10
Estimated value	800	500	400	300	200	100	80	60	40	20
Exponential curve	871	584	382	262	176	118	79	53	36	24

In the second year there is an allowance of 25 per cent against the current written-down value, so that the written-down value in the second year is

$$0·65C-0·25\times0·65C = 0·65\times0·75C.$$

Similarly, in year t the written-down value is

$$0·65t\times0·75^{t-1}C.$$

Thus the total allowance $L(t)$ over t years is

$$L(t) = 0·35C+0·25\times0·65C(1+0·75+0·75^2+...+0·75^{t-2})$$
$$= (1-0·65\times0·75^{t-1})C.$$

Finally, when the equipment is sold there is a balancing adjustment to bring the total allowances equal to the net capital cost. Thus if the equipment is replaced after t years the balancing allowance is

$$C-s(t)-L(t).$$

The total allowances $A(t)$ are given by the sum of the investment, initial and annual allowances together with the balancing adjustment. Hence

$$A(t) = 0·30C+L(t)+C-L(t)-s(t)$$
$$= 1·30C-s(t).$$

We will now examine what the optimal replacement period is using the two criteria of minimizing average annual costs and minimizing the present value of future costs with and without capital allowance considerations.

If x is the replacement period the average annual costs $F_1(x)$ are

$$F_1(x) = \frac{1700-1300e^{-0.4x}}{x} + \frac{1}{x}\int_0^x u(t)dt$$

$$= \frac{1700-1300e^{-0.4x}}{x} + 49.9 + 12.45x.$$

$F_1(x)$ can be minimized by a search procedure or plotted for a range of values of x, and it will be found that the minimum occurs for $x = 11$.

If capital allowances $A(x)$ are taken into account the average annual costs are

$$F_2(x) = \frac{1700-1300e^{-0.4x}-(1.30C-1300e^{-0.4x})g}{x} + \frac{1}{x}\int_0^x u(t)dt$$

where g is the rate of taxation taken to be 0.4 and C is the capital cost of £1700. This gives

$$F_2(x) = \frac{816-780e^{-0.4x}}{x} + 49.9 + 12.45x$$

and $F_2(x)$ is a minimum for $x = 7$. The capital allowances therefore cause a reduction of 4 years in the economic replacement period.

The objective of minimizing the present value of future costs will now be considered. Let i denote the rate of interest and $d = 1/(1+i)$ be the discount rate. Then if the replacement period is x the discounted value of total costs over one period $f_3(x)$ are

$$f_3(x) = 1700-1300e^{-0.4x}\ d^x + \int_0^x u(t).d^t dt.$$

Thus for an infinite series of successive replacements the total of all future costs discounted to the present is $F_3(x)$ where

$$F_3(x) = f_3(x)(1+d^x+d^{2x}+...) = \frac{f_3(x)}{1-d^x}.$$

This gives after integration

$$F_3(x) = \frac{(1700-1300e^{-0.4x})}{1-d^x} - \left(49.9-24.9\left(\frac{1}{\log d}+\frac{xd^x}{1-d^x}\right)\right)\frac{1}{\log d}.$$

$F_3(x)$ can be minimized directly for a variety of values of the interest rate. The replacement period varies from 12 to 16 years depending on the interest rate as shown in Table 6.4.

When capital allowances are included in the replacement model the objective function $F_4(x)$ has the same form as $F_3(x)$

$$F_4(x) = \frac{f_4(x)}{1-d^x}$$

and the present value of the capital allowances $P(x)$ are included in $f_4(x)$ as

$$f_4(x) = f_3(x) - P(x)g.$$

Using the same figures as before, the capital allowances for an initial capital outlay of C over x years will be

$$0{\cdot}65C \quad d + 0{\cdot}25 \times 0{\cdot}65C(d^2 + 0{\cdot}75d^3 + \dots + 0{\cdot}75^{x-2}d^x)$$

$$= 0{\cdot}65C \quad d + 0{\cdot}1625C \quad d^2(1 - (0{\cdot}75d)^{x-1})/(1 - 0{\cdot}75d).$$

The discounted value of the balancing adjustment will be

$(0{\cdot}65(0{\cdot}75^{x-1}C) - s(x))d^x$.

Table 6.4

Interest rate %		1	2	3	4	5	6	7	8	9	10	11	12	13	14	15
Optimum replacement period	Without capital allowances	12	12	12	12	13	13	13	14	14	14	15	15	15	16	16
	With capital allowances	8	8	8	8	9	9	9	9	10	10	10	11	11	11	12

The quantity $P(x)$ consists of the sum of these expressions.

$F_4(x)$ is now minimized for a variety of values of the interest rate and the results are shown in Table 6.4. The replacement period now varies from 8 to 12 years which again gives a substantial reduction on the results without capital allowances.

It can be concluded from these comparisons that, for this data, capital allowances have a significant shortening effect on the optimal replacement period for capital equipment, and that the discounting criterion tends to yield longer replacement periods than the criterion of minimizing average annual costs.

6.3 Planned renewal of a machine with technological improvement

The replacement problem is not usually as straightforward as it has appeared. It is not simply a matter of determining the intervals at which a machine should be replaced by a similar machine; the key problem is to make the replacements when significant technological improvements have occurred and better performances are offered. In this situation it may be worth while to postpone a replacement in order to take advantage of an expected technical improvement. The solution will certainly not be expressed in the simple cyclical form of Example 6.2.

The introduction of technological improvement requires a re-definition of the cost characteristics to distinguish the purchase year of the machine. We will define $r_n(j)$, $u_n(j)$ and $s_n(j)$ as the yearly revenue, operating cost and salvage value after j years of a machine installed in year n. For example, we assume that the operating costs will steadily decrease owing to technological improvement; the operating costs over a lifetime $u_n(j)$ will then become a family of curves depending on the year of manufacture of the machine as illustrated in Fig. 6.5. Given this data, and the purchase price of the machine

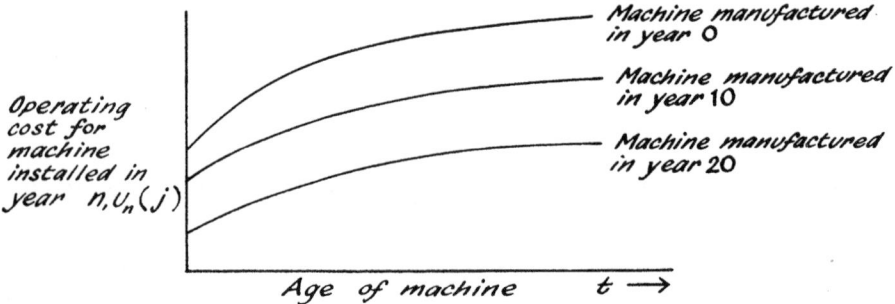

Fig. 6.5

in year n, say c_n, the problem is to determine the times at which replacements should be made so as to minimize the total discounted returns. We will consider a finite time span of T years, and assume that all replacements are made with new machines.

The control variables are the number of replacements and the years in which the replacements should be made. If N replacements are made in years $x(1)$, $x(2)$, ..., $x(N)$ we add up the costs associated with each machine. The discounted returns for the ith machine are the sum of the revenues, less the sum of the operating costs, less the difference between the cost of the new machine and the salvage value of the old one. We will refer to the last factor as the replacement cost, and will denote by $c_n(j)$ the cost of replacing a machine of age j which was installed in year n by a new machine. Thus if $y = x(i)$ and $z = x(i+1)-1$ the costs of the ith machine discounted to the first year are $g(i)$ where

$$g(i) = \sum_{t=y}^{z} (r_y(t) - u_y(t))d^{t-1} + c_y(z - y + 1)d^z.$$

The optimization problem is therefore to determine the value of the problem variables, N, $x(1)$, $x(2)$, ..., $x(N)$ to minimize the total discounted costs

$$F(N, x_1, x_2, ..., x_N) = \sum_{i=1}^{N} g(i)$$

subject to the constraints:

$$N \geq 0$$

$$x(1) \geq 0$$

$$x(i+1) \geq x(i)$$

$$x(N) \leq T$$

all variables being integer valued. This is a large and difficult optimization problem and it is fortunate that we can turn to the method of dynamic programming to solve the problem in a very elegant fashion.

In order to apply the dynamic programming procedure we need to express the calculation in terms of stages and states together with the special objective functions defining the optimal replacement policy. We consider each year to be a stage, aud the states to be the various possible ages of the machine in that year. There are therefore T stages and there are potentially T states at each stage (although it is unlikely that we will be keeping the machine until it is of age T). We define the functions $f_n(j)$ for the recurrence relations as

$f_n(j)$ = the value at year n of the overall discounted return from a machine which is j years old, where an optimal replacement policy is employed for the remainder of the process.

As the period being considered is T years, we define $f_n(j) = 0$ for $n \geq T$.

There are two possible decisions which can be made in each year (i.e. at each stage): either to keep the existing machine or to replace it with a new one. If a new machine is bought in year n to replace the existing machine say of age j, the total discounted returns from year n onwards will be the revenue $r_n(0)$ less the operating cost $u_n(0)$ of a new machine plus the replacement costs $c_{n-j}(j)$ of a machine installed in year $(n-j)$ and now of age j, plus the discounted costs from year $(n+1)$ onwards starting with a machine of age 1; i.e. the replacement returns are

$$r_n(0) - u_n(0) - c_{n-j}(j) + df_{n+1}(1).$$

If the existing machine is kept, the future discounted returns are the revenue $r_{n-j}(j)$ less the operating cost $u_{n-j}(j)$ of a machine of age j in year n plus the future costs from year $(n+1)$ onwards starting with a machine of age $(j+1)$ in year $(n+1)$; i.e. the keeping returns discounted are:

$$r_{n-j}(j) - u_{n-j}(j) + df_{n+1}(j+1).$$

As one must either replace the machine with a new one or keep the old one, and our aim is to maximize the total return, the returns at year $(n+1)$ can be related to the returns at year n by recurrence relations. We wish to replace or keep according to whichever offers the larger return, therefore

$$f_n(j) = \max \begin{cases} \text{Replace:} & r_n(0) - u_n(0) - c_{n-j}(j) + df_{n+1}(1) \\ \text{Keep:} & r_{n-j}(j) - u_{n-j}(j) + df_{n+1}(j+1). \end{cases}$$

The state j at stage n can therefore be reached from only two possible states at stage $(n+1)$, and the choice is neatly embodied in the formula for the recurrence relations. The functions $f_n(j)$ are evaluated backwards in the order $n = T, T-1, T-2, ..., 2, 1$, and for an appropriate set of values of j. As we know $f_{T+1}(j) = 0$, we can use the recurrence relations to evaluate $f_T(j)$ and then $f_{T-1}(j)$ and so on. As each decision is made the policy used in the maximization (either 'keep' or 'replace') will be recorded. If the initial age of the current machine in year 1 is j_0 the optimum will be obtained as $f_1(j_0)$ and the optimal policy can be evaluated by a forwards pass over the years $n = 1, 2, ..., T$. The procedure is illustrated in the next example reported and evaluated by Dreyfus.

Example 6.3

We will study a ten-year process. It is assumed that we currently have a machine (called the incumbent machine) which is three years old with the cost characteristics detailed in Table 6.5. For computational simplicity we

Table 6.5

Incumbent machine

Age of machine j	3	4	5	6	7	8	9	10	11	12
Revenue $r_0(j)$	60	60	50	50	50	40	40	40	30	30
Operating cost $u_0(j)$	55	55	55	60	60	60	60	65	65	70
Replacement cost $c_0(j)$	250	260	270	280	280	290	290	300	300	310

shall assume the discount rate is unity, although in fact d should be taken small enough to reflect the uncertainty of the future as well as the discounting value. Revenue, operating cost and replacement cost for machines manufactured in each of ten years are expressed in the following functions of machine age.

The dynamic programming recurrence relations can now be evaluated using the data of Table 6.6. First we compute a table of $f_{10}(j)$, the value from year 10 and all remaining years (zero in this case) for a machine of age j. We need to consider values of j corresponding to replacing a machine in any of the years 1, 2, ..., 9 giving current ages 9, 8, 7, 6, ..., 1 or the possibility that the incumbent machine is still there giving an age of 12. To compute $f_{10}(1)$ we compare the revenue minus the operating cost for the old machine (a machine made in year 9 and now one year old) to the cost of replacing. Hence

$$f_{10}(1) = \max \left[\begin{array}{ll} \text{Replace:} & r_{10}(0) - u_{10}(0) - c_9(1) = 155 - 5 - 225 \\ \text{Keep:} & r_9(1) - u_9(1) \qquad\qquad = 140 - 10 \end{array} \right\} = 130$$

Table 6.6

Machine made in year 1

Age of machine j	0	1	2	3	4	5	6	7	8	9
Revenue $r_1(j)$	90	85	80	75	70	70	70	60	60	60
Operating cost $u_1(j)$	20	20	25	25	30	30	35	40	45	50
Replacement cost $c_1(j$	200	220	240	250	255	260	265	270	270	270

Machine made in year 2

Age of machine j	0	1	2	3	4	5	6	7	8
Revenue $r_2(j)$	100	90	80	75	70	65	65	65	65
Operating cost $u_2(j)$	15	20	20	25	25	30	30	35	35
Replacement cost $c_2(j)$	200	220	240	250	255	260	265	270	270

Machine made in year 3

Age of machine j	0	1	2	3	4	5	6	7
Revenue $r_3(j)$	110	105	100	95	90	80	70	60
Operating cost $u_3(j)$	15	15	20	20	25	25	30	30
Replacement cost $c_3(j)$	200	220	240	250	255	260	265	270

Machine made in year 4

Age of machine j	0	1	2	3	4	5	6
Revenue $r_4(j)$	115	110	100	90	80	70	60
Operating cost $u_4(j)$	15	15	20	20	25	25	30
Replacement cost $c_4(j)$	210	215	220	225	230	235	240

Machine made in year 5

Age of machine j	0	1	2	3	4	5
Revenue $r_5(j)$	120	115	115	110	105	100
Operating cost $u_5(j)$	10	10	15	15	20	20
Replacement cost $c_5(j)$	210	215	220	225	230	235

Machine made in year 6

Age of machine j	0	1	2	3	4
Revenue $r_6(j)$	125	120	110	105	100
Operating cost $u_6(j)$	10	10	10	15	15
Replacement cost $c_6(j)$	210	220	230	240	250

Machine made in year 7

Age of machine j	0	1	2	3
Revenue $r_7(j)$	135	125	110	105
Operating cost $u_7(j)$	10	10	10	10
Replacement cost $c_7(j)$	210	220	230	240

Machine made in year 8

Age of machine j	0	1	2
Revenue $r_8(j)$	140	135	125
Operating cost $u_8(j)$	5	10	10
Replacement cost $c_8(j)$	220	230	240

Machine made in year 9

Age of machine j	0	1
Revenue $r_9(j)$	150	140
Operating cost $u_9(j)$	5	10
Replacement cost $c_9(j)$	220	225

Machine made in year 10

Age of machine j	0
Revenue $r_{10}(j)$	155
Operating cost $u_{10}(j)$	5
Replacement cost $c_{10}(j)$	220

and we keep the machine. (It would be surprising if it was advantageous to buy a new machine during the last stage of a fixed period.) Similarly we compute $f_{10}(2)$ by comparing the cost of keeping a two-year-old machine (made in year 8) with the replacement cost:

$$f_{10}(2) = \max \begin{bmatrix} \text{Replace:} & r_{10}(0) - u_{10}(0) - c_8(2) = 155 - 5 - 240 \\ \text{Keep:} & r_8(2) - u_8(2) \qquad\qquad = 125 - 10 \end{bmatrix} = 115$$

and again the machine is kept. We complete the table of values of $f_{10}(j)$ as shown in Table 6.7 denoting the policy to replace by (R) and the policy to keep the machine by (K).

Table 6.7

t	$f_{10}(j)$	Policy
1	130	K
2	115	K
3	95	K
4	85	K
5	80	K
6	30	K
7	30	K
8	30	K
9	10	K
12	−40	K

We now compute the table of values corresponding to $f_9(j)$. Here the values of $f_{10}(j)$ are needed through the recurrence relation

$$f_9(j) = \max \begin{cases} \text{Replace:} & r_9(0) - u_9(0) - c_{9-j}(j) + f_{10}(1) \\ \text{Keep:} & r_{9-j}(j) - u_{9-j}(j) + f_{10}(j+1). \end{cases}$$

The first two values for $j = 1$ and 2 are calculated as

$$f_9(1) = \max \begin{bmatrix} \text{Replace:} & 150 - 5 - 230 + 130 \\ \text{Keep:} & 135 - 10 + 115 \end{bmatrix} = 240$$

$$f_9(2) = \max \begin{bmatrix} \text{Replace:} & 150 - 5 - 230 + 130 \\ \text{Keep:} & 110 - 10 + 95 \end{bmatrix} = 195$$

and in both instances the machine is kept.

We can thus complete the total set of tables given in Table 6.8.

The value of $f_1(j)$ is calculated only for $j = 3$ as the initial machine was assumed to be three years old. The value of $f_1(3)$ is the total return obtainable by using an optimal policy. This quantity is calculated by comparing the possibilities of replacing or keeping the incumbent machine. If we replace the three-year-old machine at a cost of £250, the new machine manufactured in year 1 will produce a net revenue of £70 for its first year plus future profits of $f_2(1) = 490$. Therefore the net gain is $490 - 250 + 70 = 310$. If we keep the machine we accumulate a net revenue of £5 from the four-year-old initial machine and can earn $f_2(4) = £285$ in all future years. As 310 is greater than 290, we choose to purchase the new machine.

After deciding what to do in the first year we can now trace back through Table 6.8 to see what is the rest of the policy. Once we have decided to purchase in year 1, we see that we will start year 2 with a one-year-old machine, and the policy associated with $f_2(1)$ is to keep the machine for a

Table 6.8

j	$f_9(j)$	Policy	j	$f_8(j)$	Policy	j	$f_7(j)$	Policy
1	240	K	1	310	K	1	385	K
2	195	K	2	275	K	2	360	K
3	175	K	3	260	K	3	215	K
4	165	K	4	145	R	4	190	K
5	75	K	5	125	K	5	175	R
6	70	K	6	110	R	6	170	R
7	60	K	7	105	R	9	145	R
8	25	K	10	75	R			
11	−25	R						

j	$f_6(j)$	Policy	j	$f_5(j)$	Policy	j	$f_4(j)$	Policy
1	465	K	1	390	K	1	435	K
2	295	K	2	345	K	2	385	K
3	265	K	3	325	R	3	370	K
4	245	R	4	320	R	6	285	K
5	240	R	7	295	R			
8	210	R						

j	$f_3(j)$	Policy	j	$f_2(j)$	Policy	j	$f_1(j)$	Policy
1	455	K	1	490	K	3	310	R
2	425	K	4	285	K			
5	280	K						

further year. We start year 3 with a two-year-old machine and the policy for $f_3(2)$ is to keep the machine for a further year. Retracing our steps in this manner, we see that the optimal policy is to purchase in year 1, keep until year 5, when we purchase again, and then keep this machine for the rest of the period. As a check on the calculations, we can add up the gain from this policy, which is shown in Table 6.9 and it will be found that the total profit = 310.

This analysis includes some interesting by-products. Should we be forced to keep the incumbent machine during year 1, perhaps because of insufficient capital to replace, our optimal policy then says that we should keep the old machine until year 5 and then replace. This is the policy that results in a net profit of £290. Since the net gain of replacing now over that of keeping is £20 it is for management to decide if the initial outlay of £250 for a £20 net gain is economical.

Table 6.9

Year	Policy	Profit
1	R	−180
2	K	65
3	K	55
4	K	50
5	R	−145
6	K	105
7	K	100
8	K	95
9	K	85
10	K	80

The dynamic programming technique can be extended to more general equipment replacement problems. It was assumed that the replacement would always be made with a new machine. If a partially used machine can be bought this must be considered as an alternative decision. The recurrence relations which were used for the previous example would now have to be revised to include this possibility. Let $g_{n-j}(j, x)$ denote the cost of replacing a machine of age j in year n by a machine of age x. Then defining $f_n(j)$ in the same way as before, the recurrence relations become:

$$f_n(j) = \max \begin{cases} \text{Purchase new machine:} \\ r_n(0) - u_n(0) - c_{n-j}(j) + df_{n+1}(1). \\[2mm] \text{Purchase a machine of age } x: \\ \max_{x \geq 1} r_{n-x}(x) - u_{n-x}(x) - g_{n-j}(j, x) + df_{n+1}(x+1) \\[2mm] \text{Keep the old machine:} \\ r_{n-j}(j) - u_{n-j}(j) + df_{n+1}(j+1). \end{cases}$$

Dreyfus also discusses how dynamic programming may be used to include the possibility of an overhaul to a machine where the overhaul effectively reduces the age of the machine. The recurrence relations can also accommodate some practical restrictions such as a limitation that cash balances should not be allowed to fluctuate unduly in the course of the replacement process. The dynamic programming procedure is an outstanding success on this type of problem.

6.4 Capacity renewal in an industry: the electricity generating system

So far we have studied the replacement of one machine by another without regard to capacity considerations. In practice a manufacturer will usually consider the productivity of new machines on the market when making renewal decisions. He may decide that a purchase of a new machine will offer 50 per cent more output than the old machine and he will have to assess whether he can sell the additional product. The replacement decision will thus depend on market conditions.

Plant investment often takes place in periods of market expansion and the purchase of plant therefore represents both renewal decisions and expansionary investments. The problem becomes particularly difficult in large industries subject to a seasonal demand pattern, as then, some plants may be operated over long periods and some over short periods. In this situation, even when it has been decided that new capacity is needed, it may not be clear which is the best new type of plant to purchase as its operation policy will affect the operating policies for the existing plants. If the new equipment is automatic, faster and is inexpensive to run it may mean that some of the existing plants are not fully utilized. It might therefore be better to purchase cheaper equipment for peak-load operation. This kind of plant renewal problem arises in large chemical industries where there is a seasonal demand, but the classical case occurs in the electricity generating system where the seasonal demand pattern cannot be smoothed by storage of the product. The next example considers this problem in detail.

Example 6.4   Plant investment in the electricity generating system
The demand for electricity varies substantially in the course of a year reaching a peak in winter and a low point in summer. Fig. 6.6 gives an illustrative

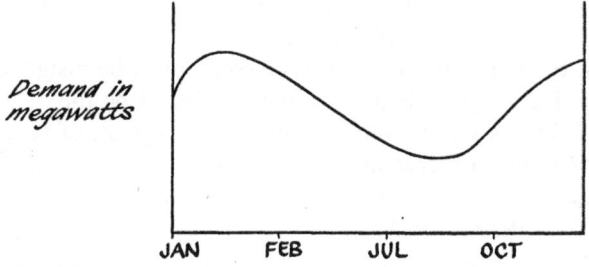

Fig. 6.6

curve of the daily demand in the course of a year. As electricity is not storable, the supply must be generated as required and therefore the plant capacity must equal the maximum daily demand which occurs in the course of a year. The capacity requirements can therefore be plotted over the years as the increasing maximum demand as shown in Fig. 6.7.

The consequence of non-storeability and the fluctuating demand is that some plants will have to operate over long periods and some over short periods. Furthermore, in the technologies of electricity generating plant it is generally true that plants either have high capital costs and low fuel costs or low capital costs and high running costs. Nuclear and hydro-electric power fall into the former category, coal and oil-fired stations into the latter. Low fuel costs are needed for the base load operating throughout the year, and low capital costs are required for plants which only operate for short periods. Thus there is no single best plant type for installation purposes, and the optimal plant investment policy will normally include a mixture of plant types.

We will now formulate the optimization problem in detail. The planning objective for the electricity generating system is to choose the new installations so as to minimize the total discounted system costs over a given period of

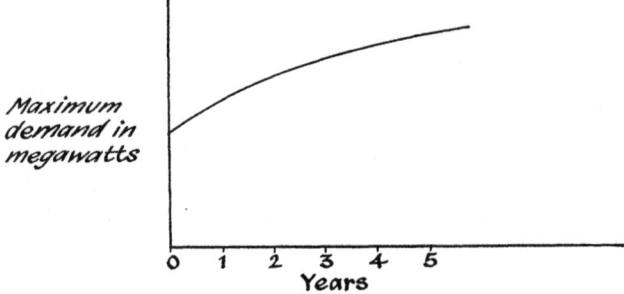

Fig. 6.7

years. It is assumed that the demand figures have been forecast both for the increasing maximum demand and the seasonal variation. It is also assumed that the generating system is fully integrated and that transmission costs are not considered. The installation decisions will take place annually and all new plants will come on-power at the beginning of the year.

The problem variables are the capacities of each type of plant to be installed in each year. Let T be the period of years being studied and let M be the number of plant types. Let $L(k)$ denote the planned lifetime of a plant of type k. We will then denote the capacity in megawatts of plant type k to be installed in year t by $x(k, t)$ for $k = 1$ to M and $t = 1$ to T, and the total set of $x(k, t)$ variables will be summarized by the vector X,

$$X = (x(1, 1), x(2, 1), x(3, 1), ..., x(M, 1), x(1, 2), ..., x(M, T)).$$

We now define the objective function expressing the total discounted system costs. This is quite a complicated task. The cost associated with any individual plant is its capital cost plus the product of the fuel cost and its number of hours of operation in each year summed over its lifetime of operation. Let $a(k, t)$ denote the capital cost in £/megawatt of a plant of type k installed in year t and let $r(k, t, i)$ denote its fuel cost in £/mw/hr in year

$i(i \geq t)$. Also let $h(k, t, i)$ denote the number of hours of operation in year i of a plant of type k installed in year t. (It is assumed that all capacity of the same type installed in the same year operates for the same number of hours.) The running cost in year i of 1 megawatt of plant of type k installed in year t is therefore

$$r(k, t, i) . h(k, t, i)$$

and the total cost of one megawatt installed in year t over the whole period is

$$a(k, t) + \sum_{i=t}^{T(k, t)} r(k, t, i) . h(k, t, i)$$

where the upper limit of the summation is either the end of the period T or the end of the life-time $t + L(k)$, so that

$$T(k, t) = \min (T, t + L(k)).$$

Therefore if $x(k, t)$ megawatts of plant type k are installed in year t the total cost for this installation is the product of $x(k, t)$ and the last expression. The total discounted system costs are the sum of the costs of the plants of all types $(1 \leq k \leq M)$ installed in all years $(1 \leq t \leq T)$ discounted to year 1. Therefore if $d(t)$ is the discount rate in year t the formula for the discounted system costs is

$$F(X) = \sum_{t=1}^{T} \sum_{k=1}^{M} x(k, t) \left(a(k, t) + \sum_{i=t}^{T(k, t)} r(k, t, i) . h(k, t, i) \right) \bigg/ \Pi_{j=1}^{t} (1 + d(j)).$$

It should be noted that some of the capital costs may have to be adjusted in this formula; it would be incorrect to include the full capital costs for a plant installed in year T. Therefore, towards the end of the period the capital costs will be taken as that proportion of capital costs which fall within the relevant period, the proportion referring to the typical lifetime of a plant. If $A(k, t)$ denotes the total capital costs of one megawatt of a plant of type k installed in year t, the $a(k, t)$ values are determined as

$$a(k, t) = A(k, t) \text{ for } 1 \leq t \leq T - L(k)$$

$$a(k, t) = \frac{(T - t + 1)}{L(k)} A(k, t) \text{ for } T - L(k) < t < T.$$

Although the objective function formula may look linear and comparatively straight forward, it is in fact non-linear as the operating variables $h(k, t, i)$ depend on the other $x(k, t)$ values. This is the principal mathematical difficulty. The $h(k, t, i)$ quantities depend on how the plants are going to be operated collectively. Because the demand for electricity varies over the year some plants will be required to operate over long periods and others over short periods. The distribution of operating durations is defined by the plant load curve which gives the spectrum of operating requirements. A typical curve is shown in Fig. 6.8. Clearly we want to run the cheap fuel plants for long periods and the expensive fuel plants for short periods. Therefore we arrange the plants in the system in any year in a 'merit order' corresponding to increasing fuel cost. The plants are now placed in the merit order along the base of the load curve and the ordinates of the load curve give the number of hours of operation. Fig. 6.8 shows how a plant in position p of the merit order performs $H(p)$ hours of operation. The operating hours of a particular plant therefore depend on how much of each type of plant is currently in the system and thus the $h(k, t, i)$ values depend on all the $x(k, t)$ variables.

In suitable circumstances we can express $h(k, t, i)$ in an analytic form. Suppose the indices k are arranged in order of increasing fuel cost, i.e. $r(k, t, i) < r(k+1, t, i)$. Also let us assume that steady technological improvement occurs, so that the running costs steadily fall. In this case we know that

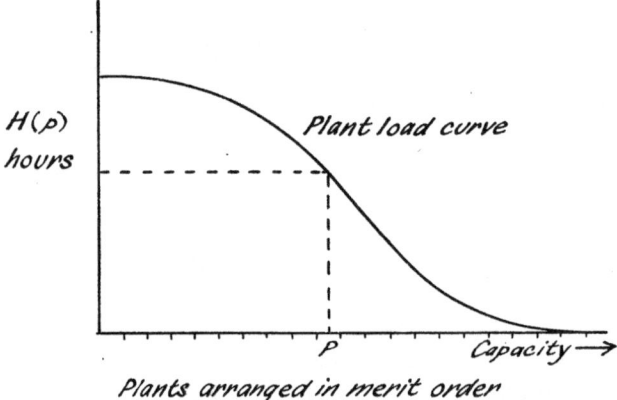

Fig. 6.8

the merit order is arranged in order of increasing k index, and, for a given k index, in order of increasing age of installation. Thus the total plant capacity $p(k, t, i)$ operating in year i at a cheaper cost than a plant of type k installed in year t consists of all plants of types 1, 2, ..., $(k-1)$, installed up to year i which have not yet retired plus plants of type k installed since year t:

$$p(k, t, i) = \sum^{k-1} \sum^{i} x(l, j) + \sum_{j=t+1}^{i} x(k, j) + \frac{x(k, t)}{2}$$

1st $\sum$: $l = 1$ is lower limit
2nd $\sum$: $j = \max(1, t - L(l))$ is lower limit

where the last term includes half the plant of this particular age and type. If we now define $H_t(p)$ as the plant load curve for year t giving the number of hours of operation of a plant for which there is a capacity p at a cheaper running cost in year t then,

$$h(k, t, i) = H_t(p(k, t, i)).$$

The original objective function can now be completed with the function

$H_t(p(k, t, i))$ replacing $h(k, t, i)$, in the formula for $F(X)$ for the total costs. This means that $F(X)$ is now a non-linear function.

We now turn to the constraints which may be imposed on the $x(k, t)$ values. First, the planning task will usually relate to an existing generating system, so that the first say L years of the period of study will already have fixed installations. If we denote the number of plants of type k installed in year t, $(t \leq L)$ by $G(k, t)$ we set

$x(k, t) = G(k, t)$ for $k = 1, M$ and $t = 1, L$.

(The optimization study therefore relates the years $t = L+1, 2+2, ..., T$). Secondly, the total capacity must be able to meet the maximum demand and therefore

$$\sum_{k=1}^{M} \sum_{l=1}^{t} x(k, l) \geq D(t),$$

where $D(t)$ denotes the maximum demand for electricity in year t, and an appropriate adjustment is made in the formula for cases where plants have already been retired.

Thirdly, all the $x(k, t)$ values must be non-negative, i.e.

$x(k, t) \geq 0.$

Fourthly, we may need to impose the practical constraint that installation rates must not be allowed to fluctuate excessively. We therefore impose the smoothing constraints of the form

$x(k, t-1) - B(k) \leq x(k, t) \leq x(k, t-1) + B(k)$

where $B(k)$ is the maximum permissible change in the rate of installation of plant type k from any one year to the next.

Fifthly, to achieve the capital cost figures $a(k, t)$, it may be necessary to achieve a minimum rate of installation. We denote this minimum rate of installation in year t for a plant of type k by $m(k, t)$ so that

$x(k, t) \geq m(k, t).$

We have thus constructed an optimization problem in which the objective function is non-linear and the constraints are all linear inequalities. The projected gradient method of non-linear programming can now be used to calculate the minimum.

The method was applied to a case study which considered two types of plant which are distinguished as nuclear and conventional power, the latter being coal- or oil-fired generating stations. Three studies were made over 5, 10 and 15 years respectively. The capital and fuel cost characteristics of a plant to be installed in any year are shown in Figs 6.9 and 6.10. It is assumed that the fuel costs remain constant over a plant's lifetime.

The percentage increase in the maximum demand for electricity per year for the period is assumed to be as shown in Fig. 6.11, which leads to the maximum demand curve of Fig. 6.12.

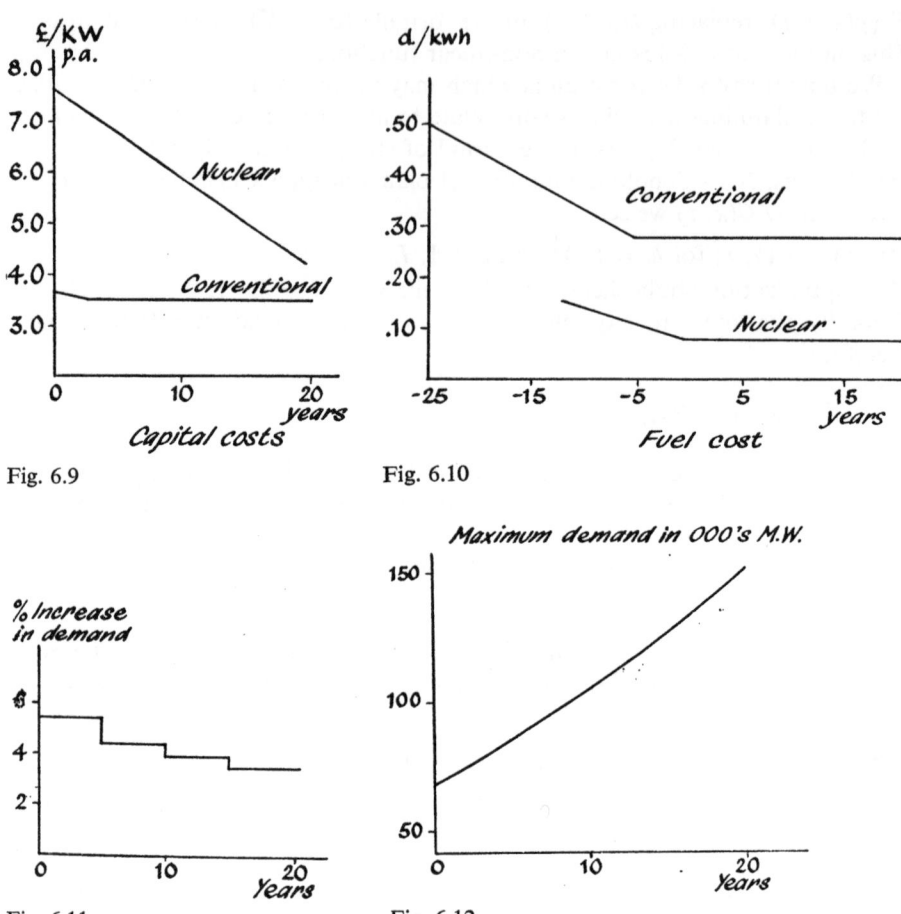

Fig. 6.9

Fig. 6.10

Fig. 6.11

Fig. 6.12

The lifetime of each plant type is assumed to be 25 years. The discount rate is 8 per cent over the whole period.

Table 6.10

Year	Installation values (m.w.'s)	
	Nuclear	Conventional
1	4934	0
2	5532	0
3	5741	0
4	5961	0
5	5383	0

The results for the 5, 10 and 15 years' studies are shown in Tables 6.10, 6.11 and 6.12.

Table 6.11

10-year period

| Year | Installation values (m.w.s) | |
	Nuclear	Conventional
1	4934	0
2	4895	637
3	3760	1981
4	4240	1721
5	4453	930
6	5441	200
7	6010	0
8	5960	0
9	5686	0
10	5557	0

Table 6.12

15-year period

| Year | Installation values (m.w.s) | |
	Nuclear	Conventional
1	4934	0
2	1114	4418
3	3218	2523
4	3616	2345
5	3777	1606
6	4311	1330
7	6010	0
8	5960	0
9	5686	0
10	5557	0
11	5868	0
12	6187	0
13	6662	0
14	6669	0
15	6102	0

The tables demonstrate how the optimum policies will differ according to the length of the period of optimization. If optimization is carried out over a 5-year period, Table 6.10 shows how the installation is all nuclear. On the

F

other hand, if the system costs are to be minimized over 15 years (Table 6.12), then in the first 5 years over 40 per cent of the installation are conventional plants. The long-term and short-term studies determine different policies in years 1 to 5, because the long-term optimization takes account of predicted future cost patterns. The cost of nuclear power is assumed to fall steadily and significantly (and by a greater amount than conventional power) over the whole period, so that the policy indicated by long-term optimization is of some postponement of the installation of nuclear power until it is really cheap. It is clear, therefore, that the optimization process is critically dependent on the assumptions about future levels of cost. If the optimization procedure were used as aid to management decisions, a range of forecasts would have to be made about demand and cost characteristics with the final decision being based on an analysis of the probabilities and acceptabilities of alternative strategies.

6.5 Plant location problems: choosing between pre-selected sites

When expansionary investments are being undertaken the problem of siting the plant in a good position in relation to the market may be as important a decision as the choice of plant and its renewal timing. We will examine some general forms for the plant location problem; first, we consider which of a number of given points should be chosen for the plants to minimize set-up, production and distribution costs, and secondly we will discuss a heuristic method for choosing the possible sites in relation to the market.

Suppose there are N distinct points to be considered for locating a plant (or a number of plants) as sources of supply of a single product. These locations are numbered 1, 2, ..., N. There are M distinct market points to be served, numbered 1, 2, ..., M. There is a constant annual rate of requirements R_j, to be satisfied at market j. The costs of satisfying these requirements are composed of the set-up costs of establishing a plant on a site, manufacturing costs at the various plant locations plus transport costs to individual markets. Transportation costs between each pair of demand and supply points are proportional to the amount shipped and to the distance between the points and manufacturing costs of a plant are proportional to total output.

Two sets of problem variables are needed to define where the plants should be located and how much should be manufactured at each plant for each market. Let the variable y_i be set to one or zero according to whether a plant is or is not located at point i, and let x_{ij} denote the annual rate of manufacturing at plant location i for distribution to market destination j.

We can now construct the objective function of expressing total costs for the set-up and one year's production and distribution costs. (In practice we will normally consider the production and distribution costs for a period of years or else take only a proportion of the capital costs as set-up costs.) Let t_{ij} denote the unit transportation cost from plant site i to market point j.

Let c_i be the unit manufacturing cost at site i and let a_i be the fixed charge associated with the construction of a plant at location i.

The total system costs are then

$$\sum_{i=1}^{N} a_i y_i + \sum_{i=1}^{N} \sum_{j=1}^{M} (c_i + t_{ij}) x_{ij}.$$

The constraints require that the market requirements R_j be satisfied,

$$\sum_{i=1}^{N} x_{ij} = R_j \text{ for } j = 1, ..., M.$$

Also the x_{ij} and y_i variables are related in a conditional way. If no plant is located at site i, then $y_i = 0$ and all production at site i must be zero, i.e. $x_{ij} = 0$ for all j. Only if $y_i = 1$ can the x_{ij} be non-zero; this gives the set of conditional constraints.

$$\left. \begin{array}{l} x_{ij} = 0 \text{ if } y_i = 0 \\ x_{ij} \geq 0 \text{ if } y_i = 1 \end{array} \right\} \text{ for } i = 1, ..., N, \text{ and for all values of } j.$$

Finally, the y_i variables are 0 or 1,

$$0 \leq y_i \leq 1, \, y_i \text{ integral}.$$

Before discussing how this problem can be solved let us consider a small example which can be solved by inspection.

Example 6.5

Suppose there are four markets to be served, and that the annual demand is 10 units in each of these markets. The markets are arranged geographically as shown in Fig. 6.14, and each of the markets is a potential location for a

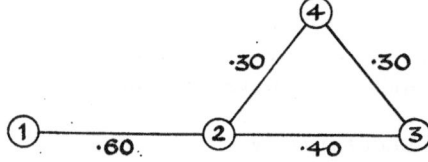

Fig. 6.13

plant. The transport costs are marked on the diagram, and all shipments to or from point 1 must pass through point 2. The transportation costs are therefore $t_{13} = t_{31} = 1{\cdot}00$ and $t_{14} = t_{41} = 0{\cdot}90$. We assume that the manufacturing costs are $0{\cdot}5$ for each unit at each plant and the plant set-up costs are $3{\cdot}5$. As the demand is going to be satisfied, the manufacturing costs will be 20 regardless of plant location and they may be ignored for optimization purposes.

There are four y_i variables each of which may assume the values 1 or 0. Therefore there are $2^4 = 16$ possible solutions in terms of the y_i variables. (We need only consider 15 as the solution $(0, 0, 0, 0)$ is not feasible.) For example the solution (1, 1, 0, 0) denotes that $y_1 = y_2 = 1$ and $y_3 = y_4 = 0$ and plants have been set up at locations 1 and 2. In this case it can be seen that the minimum transportation costs are attained by utilizing plant 1 to supply market 1 and plant 2 to supply markets, 2, 3 and 4. Total transportation costs then amount to $0.3 \times 10 + 0.4 \times 10 = 7$. The total set-up costs are

Table 6.13

All possible solutions

Solution				Plant set-up costs	Transportation costs	Total costs
y_1	y_2	y_3	y_4			
1	0	0	0	3·5	25·0	28·5
0	1	0	0	3·5	13·0	16·5
0	0	1	0	3·5	17·0	20·5
0	0	0	1	3·5	15·0	18·5
1	1	0	0	7·0	7·0	14·0
1	0	1	0	7·0	7·0	14·0
1	0	0	1	7·0	6·0	13·0
0	1	1	0	7·0	9·0	16·0
0	1	0	1	7·0	9·0	16·0
0	0	1	1	7·0	12·0	19·0
1	1	1	0	10·5	3·0	13·5
1	1	0	1	10·5	3·0	13·5
1	0	1	1	10·5	3·0	13·5
0	1	1	1	10·5	6·0	16·5
1	1	1	1	14·0	0	14·0

7·0. We can similarly calculate by inspection the costs associated with any combination of plant locations, and the costs associated with the 15 possibilities are listed below. It will be seen from Table 6.13 that the minimum possible cost of 13·0 occurs for plants located at positions 1 and 4.

In general, it is computationally too expensive to solve plant location problems by a total enumeration of all possibilities. However, we can use

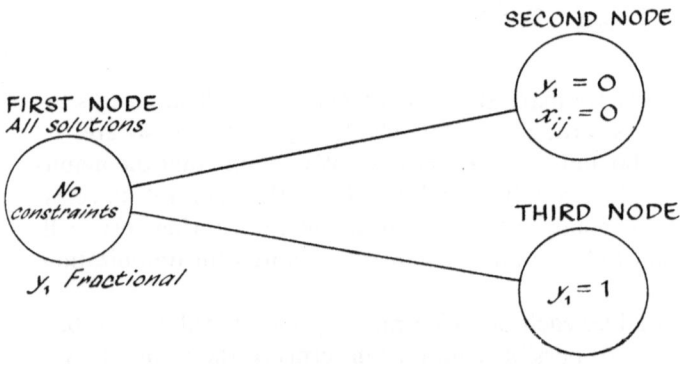

Fig. 6.14

the branch and bound technique combined with linear programming in the manner described for integer-linear programming in Chapter 8. The tree structure of the branch and bound method accommodates the conditional constraints very efficiently. The problem is first solved by the simplex method disregarding the integer constraints on the y_i values. If all the y_i values are zero or unity the optimum has been found. If however a y_i value, say y_1 is fractional the problem is partitioned into two sub-problems with $y_1 = 0$ or $y_1 = 1$. For the sub-problem associated with $y_1 = 0$ all x_{1j} values are also zero, and the new linear programming problem is solved for this commitment on this branch. The two partial solutions are shown in Fig. 6.14.

We will now review a larger plant location problem applied to the synthetic fertilizer industry reported by Vietorisz and Manne.

Example 6.6

Vietorisz and Manne have studied the plant location problem in connection with the supply of fertilizer to the South American market. The plants to be located consist of two types: primary plants producing ammonia, and secondary production units producing nitrate and sulphate fertilizer. The primary plant forms the feed to the secondary plant. The fertilizer product is then distributed to the market requirements. We need to decide how to locate both ammonia plants and the fertilizer plants and the joint distribution policies.

The following notation will be adopted:

N = number of possible sites for the ammonia plants

M = number of possible sites for the fertilizer plants

L = number of market locations

a_i = fixed annual charge for the construction and operation of an ammonia plant at location i

b_j = fixed annual charge for the construction and operation of a fertilizer plant at location j

c_{ij} = variable annual construction and operating cost per unit of annual ammonia production at i plus the cost of transportation to j

d_{jk} = variable annual construction and operating cost per unit of annual fertilizer production at j plus the cost of transportation from j to market k

R_k = the annual requirements of fertilizer at market k

x_{ij} = units of ammonia produced at location i and shipped to fertilizer production location j

y_{jk} = units of fertilizer produced at location j and shipped to market k

w_i = 1 or 0 according to whether an ammonia plant is or is not located at site i

z_j = 1 or 0 according to whether a fertilizer plant is or is not located at site j

The problem variables are x_{ij}, y_{jk}, w_i and z_j. The total annual costs to be minimized are

$$\sum_{i=1}^{N} a_i w_i + \sum_{j=1}^{M} b_j z_j + \sum_{i=1}^{N} \sum_{j=1}^{M} c_{ij} x_{ij} + \sum_{j=1}^{M} \sum_{k=1}^{L} d_{jk} y_{jk}.$$

The constraints require that the total ammonia must go into fertilizer production:

$$\sum_{i=1}^{N} x_{ij} = \sum_{k=1}^{M} y_{jk} \text{ for all values of } j.$$

The market requirements must be satisfied.

$$\sum_{j=1}^{M} y_{jk} = R_k \text{ for all } k.$$

Finally production can only take place where a plant has been established; the following conditional constraints must therefore be enforced.

if $w_i = 1$, $x_{ij} \geq 0$ for all j

if $w_i = 0$, $x_{ij} = 0$ for all j

if $z_j = 1$, $y_{jk} \geq 0$ for all k

if $z_j = 0$, $y_{jk} = 0$ for all k.

Vietorisz and Manne solved this problem for 15 possible locations for both ammonia and fertilizer plants and they considered 12 markets. Their study forms an interesting case in plant location and in the reference a full consideration is given to the way in which the cost factors have been devised for the practical situation.

*6.6 Plant location without pre-selected sites

Sometimes the plant sites are not decided in advance and it is necessary to choose the appropriate centres for possible plant location. It is presumed that a market demand distribution is known and it is required to select a given number of plant locations or supply points so that, for these supply points, the distribution costs will be minimized. Mathematically this is quite a difficult problem and we will consider it simply for the location of two sites.

Suppose two warehouses are to be located to supply N markets. The markets will be considered to be concentrated at points lying in a plane, the ith market point having coordinates (a_i, b_i). The level of demand at market point i is d_i and the transport costs between a warehouse and a market are assumed to be proportional to the amount of product carried and the distance it is moved. It is assumed that the distance between the points are the straight line planar distances and that a market point will be supplied from one warehouse only.

The problem variables are the coordinates (x_1, y_1) and (x_2, y_2) of the two warehouses and the partition of the market points into two groups for the warehouses. Suppose there are a set M_1 of markets assigned to the first warehouse and a set M_2 assigned to the second. Then, as the distance from (x_1, y_1) to (a_i, b_i) is geometrically

$$\{(x_1 - a_i)^2 + (y_1 - b_i)^2\}^{\frac{1}{2}},$$

the total costs are

$$F(M_1, M_2, x_1, y_1, x_2, y_2) = \sum_{i \in M_1} d_i\{(x_1 - a_i) + (y_i - b_i)^2\}^{\frac{1}{2}}$$
$$+ \sum_{i \in M_2} d_i\{(x_2 - a_i)^2 + (y_2 - b_i)^2\}^{\frac{1}{2}}.$$

The minimization of this function involves two parts. First, the markets have to be partitioned into two groups, and, given the groups, we need to decide where to locate the warehouse. Formally this is a difficult optimization problem. However it is possible to tackle it by a heuristic scheme.

The heuristic technique builds up two distinct market groups introducing one market point at each iteration and revising the market locations so as to minimize costs. First the two market points most widely separated are selected. These are the points $i = k(1)$, and $j = k(2)$ for which

$$(a_i - a_j)^2 + (b_i - b_j)^2$$

is a maximum. This is found by evaluation of all possibilities. $k(1)$ is assigned to the group M_1 and $k(2)$ to group M_2. Furthermore we set the current warehouse locations as

$$(x_1, y_1) = (a_{k(1)}, b_{k(1)})$$
$$(x_2, y_2) = (a_{k(2)}, b_{k(2)}).$$

From now on we proceed iteratively. At iteration t we choose the next point $k(t)$ as the point which is nearest to one or other of the current warehouse locations allocating $k(t)$ to its best group. This point is then added to the preferred group. The warehouse location of the selected group is now revised. The co-ordinates are determined so as to be at the centre of gravity of the demand distribution for the current group. If $M'_1(t)$, $M'_2(t)$ denote the groups after iteration t, the current warehouse locations $(\bar{x}_1(t), \bar{y}_1(t))$ and $(\bar{x}_2(t), \bar{y}_2(t))$ are determined by the following formulae.

$$\left(\sum_{i \in M'_1} d_i\right)\bar{x}_1(t) = \sum_{i \in M'_1} d_i a_i$$

$$\left(\sum_{i \in M'_1} d_i\right)\bar{y}_1(t) = \sum_{i \in M'_1} d_i b_i$$

$$\left(\sum_{i \in M'_2} d_i\right)\bar{x}_2(t) = \sum_{i \in M'_2} d_i a_i$$

$$\left(\sum_{i \in M'_2} d_i\right)\bar{y}_2(t) = \sum_{i \in M'_2} d_i b_i.$$

Finally it is worth noting that a special type of optimization problem arises in the petroleum and mining industries where the search for oil and mineral resources is related to the plant-siting objectives. These exploration problems are dominated by statistical and probabilistic features but some optimization studies have been carried out in these fields. Seismic reconnaissance surveys are normally conducted over a geographic region by dividing up the region into smaller cells and making a probe in each of these cells. The task of determining the optimal cell size is a statistical optimization problem. The chance of a successful probe in each cell depends on the spacing of the probes and the way in which the detectors are placed to receive the reflected signals. The neighbouring interdependence between the cells can be estimated, and then the problem of allocating a given search budget to a number of regions to be scanned simultaneously can be stated as a multistage decision process and solved by means of dynamic programming. These optimization problems in exploration for natural resources are fully discussed in the reference by Packman.

REFERENCES

Bellman, R. E. and Dreyfus, S. E. 1962. *Applied Dynamic Programming*. Princeton University Press.
Coleman, J. R., Smidt, S. and York, R. 1964. Optimum plant design for seasonal production. *Man. Sci.*, **10** (4), 778.
Dean, B. V. 1961. Replacement Theory. Chapter 8 in *Progress in Operations Research*. Ed. R. L. Aokoff. Wiley, New York.
Dreyfus, S. E. 1960. A generalized equipment replacement study. *J. Soc. Ind. Appl. Math.*, **8** (3), 425.
Efroymson, M. A. and Ray, T. L. 1966. A branch-bound algorithm for plant location. *Opns Res.*, **14** (3), 361.
Eilon, S., King. J. R. and Hutchinson. D. E. 1966. A study in equipment replacement. *Opl. Res. Quat.*, **17** (1).
Griffiths, J. C. 1966. Exploration for natural resources. *Opns Res.*, **14**, 189.
Hanssmann, F. 1968. *Operational Research for Capital Investment*. Wiley, New York.
Jonas, P. J. 1966. A computer model to determine economic performance characteristics of interconnected generating stations. *Proc. 1966 Brit. Joint Comp. Conf.*
Manne, A. S. 1964. Plant location under economics of scale-decentralization and computation. *Man. Sci.*, **11** (2), 213.
Masse, P. 1962. *Optimal Investment Decisions*. Prentice Hall, New York.
Masse, P. and Gilbrat, R. 1957. Application of linear programming to investments in the electric power industry. *Man. Sci.* **3**.
Packman, J. M. 1966. Optimization of seismic reconnaissance surveys in petroleum exploration. *Man. Sci.*, **12** (8), B312.
Sasieni, M., Yaspan, A. and Friedman, L. *Operations Research—Methods and Problems*. Wiley, New York.
Sinden, F. W. 1960. The replacement and expansion of durable equipment. *J. Soc. Ind. Appl. Math.*, **8** (8), 466.
Smith, V. L. 1961. *Investment and Production*. Harvard University Press.
Thomas, H. A. and Revelle, R. 1966. On the efficient use of High Aswan Dam for hydropower and irrigation. *Man. Sci.*, **12** (8), B296.
Vietorisz, T. and Manne, A. S. 1961. Chemical processes, plant location, and economies of scale. Chapter 6 of *Studies in Process Analysis*. Eds. A. S. Manne and H. M. Markowitz.
Wardle, F. A. 1965. Forest management and operational research: a linear programming study. *Man. Sci.*, **11** (10), B260.
Whiting, I. J. 1963. Planning for an expanding electricity supply system *Opl. Res. Quat.*, **14** (2).

Exercises on Chapter 6

1 A manufacturer is offered two machines A and B. Machine A is priced at £3000 and running costs are estimated at £800 for each of the first three years, increasing by £200 a year in the fourth and subsequent years. Machine B, which has the same capacity as A, costs £1500 but will have running costs of £1200 per year for four years, increasing by £200 thereafter. Neither machine has any scrap value. On the basis of average annual costs, determine when the machines would be replaced and which machine should be selected.

2 The cost of a machine which lasts over n years is assessed by a particular company as the operating cost increase per year plus the average capital cost per year. The operating cost increases at a constant rate u per year, and the capital costs consists of the initial outlay C less the salvage value s which is assumed to be independent of the age of the machine. Determine a formula for the optimum life time for a machine whose costs are measured in this way.

3 In Example 6.3 suppose we are starting off in year 5 with a one-year-old machine and that we want the optimal replacement policy up to year 10. Use Tables 6.7 and the 6.8 to determine the total returns and the replacement points.

4 A car owner decides at the end of each year whether to keep his present car for another year or trade it in and replace it by another of the same make, not necessarily new. He wants to adopt a replacement policy which will minimize his total costs over a six-year period. The relevant costs are given in table 6.14. A six-year-old car must be traded in. Use dynamic programming to determine what the replacement policy should be to minimize costs whatever the age of his current car.

Table 6.14

Age of car in years	Buying price (£) at beginning of year	Trade price (£) of car	Running cost in year (£)
0	700	—	150
1	500	400	175
2	250	200	200
3	150	125	250
4	100	75	375
5	50	30	450
6		10	

5 A transport company has to plan its capacity to take account of a peak load demand as well as an off-peak demand. The peak-load demands lasts for time a at intensity A. The off-peak demand lasts for time b at intensity $B(\leq A)$. Thus the demand duration curve can be drawn as shown in the diagram. Two types of equipment can be purchased to meet the demand. The capital costs of these equipments are c_1 and c_2, and the running costs are r_1 and r_2 per unit time. Formulate the optimization problem to determine how much of each equipment type to purchase and also to decide the number of hours of operation of each type as a linear program.

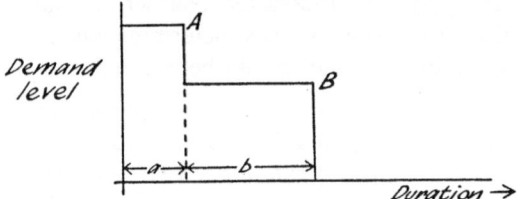

Fig. 6.15

6 A manufacturing firm is expanding its capacity to meet new market requirements for a single product. The market is to be fed from 6 warehouses located at six points which are fixed in advance. The new demand at each warehouse is given in thousands of items per annum as shown in Table 6.15.

Table 6.15

Warehouse	1	2	3	4	5	6
Demand in '000s	15	7	10	8	6	12

The new capacity to meet this demand will consist of one or more plants located at any of three possible sites. There is a different cost for setting up a plant on each of the three sites, and the cost of production is assumed to be proportional to the number of units produced. The set-up costs and cost of producing 1000 units on each site is given in Table 6.16.

Table 6.16

Plant location	1	2	3
Set up cost in £	50,000	30,000	65,000
Cost of producing 1000 units in £	400	500	300

The production capability per annum at each location is limited to 35,000 per annum on sites 1 and 2 and to 25,000 per annum on site 3.

There is a transportation cost of moving items from the plant sites to the warehouses. The cost is proportional to the quantity moved, and the cost of moving 100 items is shown in Table 6.17.

It is considered that the demand pattern will be static over the next five years. The firm wishes to plan its plant location, production and distribution over the next five-year period. Formulate the optimization problem so as to minimize the total costs and meet the demand.

Table 6.17

		To warehouse				
	1	2	3	4	5	6
From 1	27	17	35	42	16	12
plant 2	41	36	18	10	56	28
site 3	22	49	5	25	29	46

*7 Sometimes it may be necessary to determine what will be the optimal installation pattern for the electricity generating system for the next year. Suppose that there are M plant types to consider and that the set-up (or proportional capital cost) for one year of plant type k is a_k per megawatt, and that the fuel cost is r_k per megawatt per hour. Let G_k denote the amount of generating plant of type k already committed to be in the system in the year under investigation. It is further assumed that the fuel costs of the plants are ordered by increasing k index so that $r_k \leqq r_{k+1}$. The maximum demand which occurs in the next year is D, and the load curve corresponding to its distribution is defined by $H(p)$ as in Fig. 6.9. Show how dynamic programming can be used to determine the quantities of each plant to install so as to minimize total costs for the next year.

*8 A manufacturer assesses his potential market in a new region as being located at six different points. The six points can be considered to lie in a plane with a given origin for reference and coordinates in miles from the origin as shown in Table 6.18, where the demand levels are also given.

Table 6.18

Point Number	1	2	3	4	5	6
x-coordinate	0	1	4	4	5	6
y-coordinate	0	3	0	2	4	6
Demand level per week	110	80	40	140	30	100

The manufacturer wants to determine whether to establish one or two depots. The set-up cost of establishing and running a depot are estimated at £5000 per annum and the total transportation cost including vehicles and administration is estimated at 5 shillings per unit per mile. Use the heuristic technique of Section 6.5 to decide where a single depot would be located and where two depots would be located.

Then determine which scheme would offer the lower total costs.

*9 Show that in the case of no technological improvement, an infinite horizon and discounting, the dynamic programming technique will determine the same objective function and replacement policy as was obtained directly in Section 6.2 for the criterion of minimizing the present value of total discounted returns.

7 Distribution and transportation

7.1 The range of problems

There is a wide variety of problems in the distribution and transportation fields where optimization methods can be used. We will consider applications in the following four fields: (*i*) Supply-Demand Allocation; (*ii*) Routing; (*iii*) Timing; (*iv*) Traffic Control.

The first group is concerned with the efficient allocation of goods from a number of alternative sources to a variety of destinations. In the simplest form of this problem there is a fixed cost of transporting any item between a source and a destination and it is simply a question of deciding how the supplies available at the sources should be distributed without regard to routing or timing considerations. This problem was one of the first investigations of operations research and although it can be formulated in terms of linear programming there is a special method for solving the problem. We shall review it and some of its extensions in Section 7.

However, as soon as we consider the actual movement of goods from one point to another, we encounter routing problems. The transportation rarely involves a single link. Typically we wish to find the shortest or cheapest route between a source point and a destination point where there is a large number of possible routes which could be taken. These problems can be tackled by dynamic programming or by branch and bound methods. Other forms of routing problems arise where we are committed to distribute to a number of demand points and then return to the starting point. A particular case of this situation is the well-known travelling salesman who has no limit to his endurance but must complete a tour through N cities in the shortest possible distance. The same problem structure occurs in more relevant contexts where a vehicle is distributing goods, and mileage, distance and carrying capacity restrictions must be incorporated in the routing possibilities.

The third type of transportation problem is concerned with the timing constraints on a transportation fleet. The scheduling of a tanker fleet may have to consider the times at which ships are required at various ports. Also the best use of railway track by a series of trains requires a shrewd means of scheduling their sequence down the track so that their timing commitments can be satisfied. Despite the obvious importance of this latter problem, no formal optimization studies appear to have been published. The problem is known as 'train diagramming', and manual graphical procedures for solving the problem are in general use.

Finally we will study some of the simplest problems in signalled traffic control. Here we take the stand-point of how the route should deal with the traffic rather than how the transportation requirements should deal with the routes. These traffic control problems are very important in practice and are currently at an early stage of development.

7.2 A basic distribution problem

The classical transportation problem of operations research is to find the best way of allocating known supplies at a number of sources to a number of demands scattered over a variety of destinations. The cost of transportation along any link is known and it is assumed to be independent of the amount transported along the link. This problem can readily be formulated in linear programming terms.

Let M be the number of source points numbered 1 to M with supplies $s_1, s_2, ..., s_M$ and let N be the number of demand points numbered 1 to N with requirements $d_1, d_2, ..., d_N$ (where the total supply equals the total demand). Also let c_{ij} be the constant cost of despatching one unit from source i to destination j.

The problem variables are the quantities which should be transported from each source to each destination. Let x_{ij} denote the quantity to go from source i to destination j. Then the objective function is the total costs which may be expressed as the sum of the products of the amounts moved along each link and the costs of the link:

$$\sum_{i=1}^{M} \sum_{j=1}^{N} c_{ij} x_{ij}.$$

The constraints are the limitations on the supplies and demands. As the total supply is equal to the total demand, all the out-goings from source i must equal the supply s_i:

$$\sum_{j=1}^{N} x_{ij} = s_i \text{ for } i = 1, 2, ..., M.$$

Similarly, the total amount entering destination j must equal the total demanded, giving the constraint

$$\sum_{i=1}^{M} x_{ij} = d_j \text{ for } j = 1, 2, ..., N.$$

Finally all allocations must be non-negative,

$x_{ij} \geqq 0$ for all i and j.

This problem is not quite in linear programming form. The constraint equations are not linearly independent. For if we multiply the equations with s_i by $+1$ and multiply the equations with d_i by -1 and add up all right-hand sides and left-hand sides of the resulting equations we obtain zero on both sides of the final result. This shows that the equations are linearly dependent and the simplex method will not work. However, if we drop one of the equations, say the last one

$$\sum_{i=1}^{M} x_{iN} = d_N,$$

from the matrix of equations we will have an independent set and we can use the simplex method to solve the problem.

In fact there is a special method for solving this basic distribution problem which is very suitable for hand computation of the optimum. Appropriately it is called the transportation method, and it was one of the first techniques to be developed in operations research. Before outlining the method in the next section a single example of a transportation problem will be illustrated.

Example 7.1

There are three collieries S_1, S_2, S_3 supplying coal to four power-stations D_1, D_2, D_3, D_4 and the costs of transportation are given in Table 7.1.

Table 7.1

		To		
	D_1	D_2	D_3	D_4
S_1	5	9	6	8
From S_2	8	4	6	6
S_3	5	6	8	7

The supply of coal at the collieries is 20, 20 and 10 units respectively and the demand at each of the power stations is 16, 10, 10 and 14 units. Denoting by x_{ij} the amount supplied from colliery S_i to power station D_j, we wish to minimize

$$5x_{11} + 9x_{12} + 6x_{13} + 8x_{14}$$
$$+ 8x_{21} + 4x_{22} + 6x_{23} + 6x_{24}$$
$$+ 5x_{31} + 6x_{32} + 8x_{33} + 7x_{34}$$

subject to the supply constraints

$$x_{11}+x_{12}+x_{13}+x_{14} = 20$$
$$x_{21}+x_{22}+x_{23}+x_{24} = 20$$
$$x_{31}+x_{32}+x_{33}+x_{34} = 10$$

and the demand constraints

$$x_{11}+x_{21}+x_{31} = 16$$
$$x_{12}+x_{22}+x_{32} = 10$$
$$x_{13}+x_{23}+x_{33} = 10$$
$$x_{14}+x_{24}+x_{34} = 14.$$

Again it should be noted that of the seven equations, only six are independent as the sum of the first three equations less the sum of the next three equations produces the final equation. We therefore drop one of the equations to ensure linear independence for the use of the simplex method.

7.3 The transportation method

The special technique for computing the solution to the basic distribution problem will now be described in the context of Example 7.1. First it is repeated that if $(M+N-1)$ of the equality constraints are satisfied, the remaining constraint will also hold. As a result of this, the optimal allocation uses only $(M+N-1)$ of the possible routes (i.e. only $M+N-1$ of the x_{ij} variables will be non-zero).

The first stage of the procedure is to determine an initial allocation on $(M+N-1)$ basic routes. One method of doing this has already been described in the exercises in heuristic techniques of Chapter 10, Volume 1, but we will here use an alternative method which is known as the North-West corner rule. We start at the top left-hand corner of the table and allocate as much as possible to this route. This completes either the first supply row or the first demand column and this is then eliminated from consideration. The procedure is now repeated on the reduced table and then on successively smaller tables until all the quantities have been allocated.

In Example 7.1, 16 units are placed on the route (1, 1) from supply 1 to destination 1, filling up the first column. The remaining 4 units from colliery 1 are placed on route (1, 2) to destination 2, leaving 6 to be supplied by colliery 2, and so on, giving the initial basic allocation shown within the circles in Table 7.2. The unit costs have been entered in the corners of the cells in the table. (The entries for u_i and v_j are explained below.)

The next stage is to see if savings can be made by introducing loads on any of the non-basic routes. A load can be assigned to any such route if the basic routes are re-allocated. For example, assigning one unit to route (2, 1) requires taking one from route (1, 1) adding one to route (1, 2) and finally taking one from route (2, 2). The reduction in cost is $(c_{11}-c_{12}+c_{22}-c_{21})$, which is -8, so this route should not be introduced.

Similar circuits could be constructed to give the re-allocations necessary for including each of the other non-basic routes and used to find the resulting change in costs. Fortunately there is an easier approach.

Each of the unit cost coefficients c_{ij} of the basic routes can be split into a row component u_i and a column component v_j, so that $c_{ij} = u_i + v_j$. As there are $(M+N)$ u_i and v_j values and only $(M+N-1)$ basic c_{ij}'s, we arbitrarily assign $u_1 = 0$ and then work out the others. In Table 18.2, for example, $v_1 = 5, v_2 = 9$, giving $u_2 = -5$ and so $v_3 = 11, v_4 = 11$, and finally $u_3 = -4$. This kind of calculation is always made possible by the tree-like structure of the basic routes which span every row and column of the table without themselves forming circuits.

Table 7.2

Demands *j*		16	10	10	14
	$\dfrac{v_j}{u_i}$	5	9	11	11
20	0	5 (16)	9 (4) −	6 + [11]	8 [11]
20	−5	8 [0]	4 (6) +	6 (10) −	6 (4)
10	−4	5 [1]	6 [5]	8 [7]	7 (10)

(Supplies *i* along the left margin)

In constructing a circuit to introduce a new route we alternately subtract a load from a column and then have to add to the row which has been diminished always choosing basic routes and continuing until the circuit is complete. The unit saving of introducing a non-basic route (i, j) becomes

$$u_i + v_j - c_{ij},$$

all the other terms cancelling out. Using this device, there is no need to work out any actual circuits until the route to be introduced has been determined. The $u_i + v_j$ values are entered in the non-basic cells as in Table 7.2 surrounded by squares and compared with the original costs. In the example it is apparent that the biggest unit cost saving is obtained by introducing route (1, 3) (where the saving is $0 + 11 - 6 = 5$) and the corresponding re-allocation circuit is drawn as marked in Table 7.2.

The limit to the amount which can be put on this new route is given by the smallest of the quantities on routes to be diminished (shown with a minus sign on the circuit). The route (1, 2) which gives this is dropped from the basis and the new basic allocation shown in Table 7.3 is obtained by re-scheduling the 4 units previously on this route round the circuit.

Table 7.3

Demands		16	10	10	14
v_j / u_i		5	4	6	6
20	0	5 (16)_	9 [4]	6 (4)+	8 [6]
20	0	8 [5]	4 (10)	6 (6)_	6 (4)+
Supplies 10	1	5 [6]+	6 [5]	8 [7]	7 (10)_

The u_i's and v_j's are now re-calculated and the whole process is repeated until no route can be introduced at a saving. Table 7.3 gives the new u_i and v_j values and their sums, indicating an improvement by using route (3,1), and also the circuit by which this is introduced. Six units are transferred round this circuit giving the final allocation shown in Table 7.4. This is optimal as all $u_i + v_j \leqq c_{ij}$.

Table 7.4

Demands		16	10	10	14
v_j / u_i		5	5	6	7
20	0	5 (10)	9 5	6 (10)	8 7
20	−1	8 4	4 (10)	6 5	6 (10)
Supplies 10	0	5 (6)	6 5	8 6	7 (4)

The total cost for the allocation is 268 units.

7.4 Extensions of the basic distribution problem

The simple distribution problem can be generalized to more complex situations. It was assumed that the traffic between a supply point and a destination must travel direct without going to any other supply or demand point en route. However, it may be advantageous to move materials from a supply point through another intermediate supply point and on to the demand point or even through an intermediate demand point. This is known as the transhipment problem. A cost will need to be associated with transportation between two sources or between two destinations, so that if there were three sources and four destinations as in Example 7.1, there would be a 7×7 matrix of costs. If c_{ij} is the cost of sending 1 unit from source i to destination j, c'_{ij} is the cost of sending 1 unit from source i to source j and c^*_{ij} is the cost of sending 1 unit from destination i to destination j, the full cost matrix for a problem of the same dimensions as in Example 7.1 would be as shown in Table 7.5.

Table 7.5

		S_1	S_2	S_3	Destination D_1	D_2	D_3	D_4
	S_1	0	c'_{12}	c'_{13}	c_{11}	c_{12}	c_{13}	c_{14}
	S_2	c'_{21}	0	c'_{23}	c_{21}	c_{22}	c_{23}	c_{24}
	S_3	c'_{31}	c'_{32}	0	c_{31}	c_{32}	c_{33}	c_{34}
Sources								
	D_1	c_{11}	c_{21}	c_{31}	0	c^*_{12}	c^*_{13}	c^*_{14}
	D_2	c_{12}	c_{22}	c_{32}	c^*_{21}	0	c^*_{23}	c^*_{24}
	D_3	c_{13}	c_{23}	c_{33}	c^*_{31}	c^*_{32}	0	c^*_{34}
	D_4	c_{14}	c_{24}	c_{34}	c^*_{41}	c^*_{42}	c^*_{43}	0

We would now fix the artificial supplies and requirements at the sources and destinations. It would be no use setting the supplies at D_1, D_2, D_3, D_4 and the demands at S_1, S_2, S_3 to zero as no transhipment would ever occur. We therefore add to each supply point and each destination a number large enough to cover all potential transhipments. A suitable number is the total number of units originally available at S_1, S_2 and S_3. (It should be noted that we can always send the surplus units from a source to itself at zero cost.) The problem is then in the form of a standard transportation problem and can be solved by the standard method.

In addition to the possibility of directing traffic from a source to a destination via other source points or destinations, there may be restrictions on the routes which are available or the capacities which the routes can handle. We can enforce these restrictions of ensuring that traffic on these routes is limited by setting the appropriate $x_{ij} \leq C$ where C is the capacity which the route can handle.

If, however, there are numerous route restrictions, then we may not be sure that a feasible solution exists. In these circumstances we may require merely that as many of the source units are supplied as possible. We could then ignore the cost factors and present the problem as

$$\text{maximize} \sum_{i=1}^{N} \sum_{j=1}^{M} x_{ij}$$

subject to

$$\sum_{i=1}^{N} x_{ij} \leqq d_j, \text{ for } j = 1, ..., M$$

$$\sum_{j=1}^{M} x_{ij} \leqq s_i, \text{ for } i = 1, ..., N$$

$$x_{ij} \geqq 0.$$

Again, although this problem is in linear programming form some specialized procedures have been devised for its solution. A technique has been developed by Ford and Fulkerson which is based on the concept of flows through networks and is therefore suited to the cases where restrictions are present. The method is appropriate for manual calculations.

The transportation problem has been further generalized to the situation where the transported units must be carried in containers of given capacity. The demand constraints then have the form

$$\sum_{i=1}^{N} a_{ij} x_{ij} = d_j$$

where the a_{ij} values are known constants not necessarily integers. This form of the problem has also been analysed in the context of uncertain demand, where the frequency distribution of the d_j values is specified. We will examine a deterministic case of allocating aircraft to routes where this type of constraint arises.

Example 7.2

The following numbers of passengers (in hundreds) have to be transported by plane on the following routes over a year.

Table 7.6

Route	Numbers of passengers (100s)	Fare per 100 passengers
1	320	15
2	165	15
3	190	8

Planes of the available types of aircraft can carry different numbers of passengers dependent on the routes flown. The numbers of passengers (in hundreds) per year which can be carried are shown in Table 7.7.

The zeros in Table 7.7 indicate that some aircraft cannot fly on certain routes. The total number of aircraft available are 15 of type 1, 14 of type 2,

Table 7.7

Number of passengers

	Type of aircraft		
Route	1	2	3
1	20	15	0
2	18	13	10
3	0	14	8

and 18 of type 3, and the cost of an aircraft per year allocated to any route is given in the Table 7.8.

The cost of a lost passenger, i.e. one who cannot be accommodated, is equal to the lost revenue which would have been obtained from him. It is required to allocate aircraft to routes so as to minimize the total costs.

The problem variables are the number of aircraft allocated to each route and the number of passengers who cannot be carried. Let x_{ij} denote the number of aircraft of type j allocated to route i and let y_i denote the number of passengers (in hundreds) who cannot be accommodated on route i.

Table 7.8

Cost on each route

	Type of aircraft		
Route	1	2	3
1	12	13	—
2	12	13	15
3	—	11	14

The objective function to be minimized is the total cost of flights and the lost traffic, i.e.

$$12x_{11}+13x_{12}+12x_{21}+13x_{22}+15x_{23}+11x_{32}+14x_{33}+15y_1+15y_2+8y_3.$$

We have the following passenger demand constraints

$$20x_{11}+15x_{12}+y_1 = 320$$

$$18x_{21}+13x_{22}+10x_{23}+y_2 = 165$$

$$14x_{32}+8x_{33}+y_3 = 190.$$

The aircraft availability constraints (like the supply constraints in the transportation problem) are

$$x_{11}+x_{21} = 15$$

$$x_{12}+x_{22}+x_{32} = 14$$

$$x_{23}+x_{33} = 18.$$

Lastly, from Table 7.7 we have the following route restrictions $x_{31} = 0$ and $x_{13} = 0$.

This is a standard linear programming problem. The solution is displayed in Table 7.9.

Table 7.9

Aircraft type	Route		
	1	2	3
1	15	0	0
2	1·3	0	12·6
3	0	16·5	1.5

The total cost is 610½. It will be noticed that the number of aircraft allocated to certain routes is fractional. This can be interpreted to mean that some of the aircraft of that type on that route would fly at longer intervals than was assumed when the carrying capacity was established, and some aircraft will not fly exclusively on one route.

7.5 Finding the shortest route between two points in a network

Whenever routing problems are met in the transportation field we encounter the problem of finding the shortest route between two points in a network when there are a number of possible routes through the network. The network is specified by a set of junction points and a set of connections between pairs of points. A distance is associated with each possible connection. However, so far as the shortest route problem goes, we can assume that there is a theoretical connection between all pairs of points, and, where there is no physical connection, the distance along the connection is set to infinity. The network can therefore be identified by a set of N points numbered 1 to N, and a distance matrix D in which the element $d(i, j)$ records the distance from point i to point j. It is assumed that the $d(i, j)$ are positive. The distances may differ in different directions, and this is indicated by assigning different values to $d(i, j)$ and $d(j, i)$ the former referring to the direction i to j and the latter from j to i.

We have seen in Chapter 7 how the method of dynamic programming can be used to find the shortest route through a network which was organized in a series of stages, and an exercise in that chapter investigated how a more generalized network could be viewed as a structure suitable for the dynamic programming calculation. The method for the general network will be presented again here and then applied to an example. Suppose we wished to find the shortest route from point 1 to point N. We define $f_k(j)$ as the shortest route from the point 1 to point j using at most k connections or links of the network. Any route to j of at most k links passes through some other point, say i, reached by at most $(k-1)$ links, immediately prior to reaching point j. The shortest distance to such a point i is $f_{k-1}(i)$. Therefore the shortest route to j of at most k links is given by the recurrence relation

$$f_k(j) = \min_{1 \leq i \leq N} (f_{k-1}(i) + d(i, j)).$$

If we evaluate this recurrence relation for $j = 1, 2, ..., N$ and for

$$k = 1, 2, 3, ..., n,$$

we will determine the shortest route from point 1 to point N as $f_n(N)$. The number of iterations n which are necessary is such as to ensure that no reductions in the shortest routes from point 1 to any other point can take place by a further iteration, i.e. n is determined as the least integer such that:

$$f_n(j) = f_{n-1}(j), \text{ for } j = 1, ..., N.$$

Initially $f_0(j) = \infty$ for $j \neq 1$, $f_0(1) = 0$. We also record the route to be followed by noting the value of i for which the minimization occurs using the variable $q_k(j)$. It should be observed that in addition to finding the shortest route from point 1 to point N, the dynamic programming procedure has also found the shortest routes from point 1 to all other points.

Example 7.3

We wish to find the shortest route from point 1 to point 9 in the following

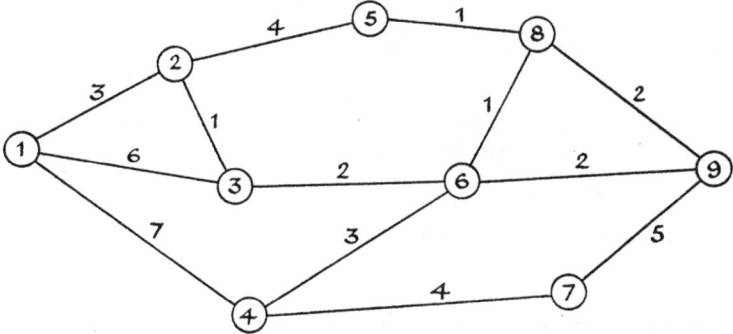

Fig. 7.1

network. The distance matrix $d(i, j)$ for this network is shown below in which a cell with no figure corresponds to an infinite distance.

Table 7.10

					To				
	1	2	3	4	5	6	7	8	9
1	0	3	6	7					
2	3	0	1		4				
3	6	1	0			2			
4	7			0		3	4		
From 5		4			0			1	
6			2	3		0		1	2
7				4			0		5
8					1	1		0	2
9						2	5	2	0

Applying the dynamic programming procedure to this problem we obtain the results tabulated below where $f_0(1) = 0$, $f_0(j) = \infty$ otherwise, and blank entries correspond to values of infinity in the $f_1(j)$ columns. For example, the

Table 7.11

j	$f_1(j), q_1(j)$		$f_2(j), q_2(j)$		$f_3(j), q_3(j)$		$f_4(j), q_4(j)$		$f_5(j), q_5(j)$	
1	0	1	0	1	0	1	0	1	0	1
2	3	1	3	1	3	1	3	1	3	1
3	6	1	4	2	4	2	4	2	4	2
4	7	1	7	1	7	1	7	1	7	1
5			7	2	7	2	7	2	7	2
6			8	3	6	3	6	3	6	3
7			11	4	11	4	11	4	11	4
8					8	5	7	6	7	6
9					10	6	8	6	8	6

value of $f_3(6)$ is calculated by using the column $f_2(j)$ and the sixth column of the $d(i, j)$ matrix for non-infinite entries as:

$$f_3(6) = \min \begin{cases} f_2(3)+d(3, 6) = 4+2 = 6 \\ f_2(4)+d(4, 6) = 7+3 = 10 \\ f_2(6)+d(6, 6) = 8+0 = 8. \end{cases}$$

The shortest route has length 8 and goes through the points $q_5(9) = 6$, $q_4(6) = 3$, $q_3(3) = 2$, $q_2(2) = 1$, i.e. it is 1, 2, 3, 6, 9.

An alternative technique for finding the shortest route between two points in a network is to use a search method of the branch and bound type. The method has been defined in its most elegant form by Dantzig. The search scheme is to 'fan out' from the starting point searching down all possible connections to adjacent points and build up the tree of routes leading out from the starting point. In this way the shortest route to each point from the starting point is determined. By a careful structuring of the calculation the distance to each point from the starting point is determined only once and this is the shortest distance.

The formal procedure of the calculation can be laid down in a few steps. At each iteration a point is selected and the shortest distance to it from the starting point is determined. The set of selected points at any iteration will be included in a set of selected points which will be denoted by S. Suppose k points have been selected. Then for any point i of the k points contained in S let

D_i = shortest distance from the starting point to i

j_i = closest point to i not contained in S

d_i = distance of point j_i from i, i.e. $d(i, j_i)$.

Then the $(k+1)$th point to be selected is j_s where

$$D_s + d_s = \min_{1 \le i \le k} (D_i + d_i).$$

This gives the minimum distance from the starting point to point j_s as $D_s + d_s$ and the route to j_s comes through the point s. The proof that this is indeed the shortest distance follows easily. Consider any other path to j_s from the origin. Eventually, the path must reach some point i not in S from a point say j in S. Provided that the distances along this path are non-negative the total distance from the starting point to j_s is not less than $D_i + d_i$. But this is at least as large as $D_s + d_s$ because of the way in which j_s has been selected. The procedure will be illustrated on a small example.

Example 7.4

Fig. 7.2 shows a network with 8 points marked by the letters 0, a, b, c, d, e, f, g. The objective is to find the shortest route from point 0 to point f. The distances between pairs of points are marked on the links.

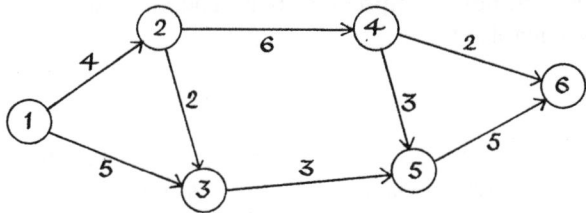

Fig. 7.2

The calculation is assisted by writing down the list of links branching out of each point in ascending order of distance. The distances are written in brackets after the links.

(0)	(a)	(b)	(c)	(d)	(e)	(f)	(g)
0–a (1)	a–b (3)	b–c (2)	c–b (2)	d–c (2)	e–f (1)	f–e (1)	g–f (1)
0–b (2)	a–c (3)	b–a (3)	c–d (2)	d–a (3)	e–c (2)	f–g (1)	g–c (3)
	a–d (3)	b–g (4)	c–a (3)	d–e (3)	e–d (3)		g–b (4)
			c–g (3)				
			c–e (3)				

Initially the set S consists of the node 0. We then proceed through the following five steps.

Step 1.   Choose link 0–a, note its distance from 0 as 1, and delete all links leading into a (i.e. 0–a, b–a, c–a, d–a). Add a to the set S.

Step 2.   Compare 0–b having length 2 with –a–b having distance $1+3 = 4$. Choose path via 0–b and record its distance from 0 as 2. Delete all links into b and include b in S.

Step 3.   Compare a–c with a–d and b–c having distances 4, 4, 4 respectively. Choose path a–c (or b–c) and a–d and record their distances from 0 as 4. Delete all links into c and d and add c and d to the set S.

Step 4.   Compare b–g, c–g and d–e with distances 6, 7, 7 respectively. Choose path b–g and record its distance from 0 as 6. Delete all links leading into g and add g to S.

Step 5.   Compare c–e, d–e, g–f with distances 7, 7, 7. Choose path via c–e (or d–e) and –g–f and record their distances from 0 as 7. Delete all links into e and f and add e and f to the set S.

All points are now included in S. The shortest path to f comes from g. The shortest path to g comes from b and the shortest path to b comes from 0. The total distance of the shortest path is 7.

It will be noted that the calculation also determines the shortest path from 0 to all other points in the network. When it is simply required to find the shortest distance from a given point to another given point, then the calculation can be reduced by starting the search from both end-points simultaneously, and fanning out until a pair of routes to a common point is found forming a through route which cannot be improved upon. The method is described in the reference by Nicholson.

7.6 Vehicle routing

Vehicle routing is a problem facing most distributors. The task is to plan the routes for the available delivery vehicles so as to make the best use of the vehicles and satisfy the market requirements. Normally the routes for each vehicle will be planned a day or two in advance and the object will be to group together the deliveries which are geographically near to one another so as to save in distribution costs. The planning is generally done on a manual basis, but it may be worth while to use an optimization technique to find the best routing plans and reveal, where possible, any surplus vehicle carrying-capacity which is available.

We will study a basic problem of determining an efficient set of routes to a known market where there may be restrictions on the distance a vehicle can go or the capacity it can carry. Let there be N customers numbered 1 to N located at specific points, which have to be supplied from a single depot. Customer i requires a quantity $q(i)$ and the distance in miles from customer i to customer j is $d(i, j)$. We regard the depot location as customer number 1. At the depot there is an unlimited supply of vehicles of capacity Q, which is the maximum total load for any route which may be formed. Also the total mileage may not exceed a stated maximum M on any route. A schedule specifies a set of say k routes $R_1, R_2, ..., R_k$, each route starting from the depot, going to a number of customers, and returning to the depot without exceeding the mileage or carrying capacity restrictions. The objective is to determine the minimum number of routes to meet the demand and such that the total mileage is a minimum.

If route R_i has n_i customers on it we can denote the customers by

$$r_i(1), r_i(2), ..., r_i(n_i).$$

The total length of route R_i will be

$$D_i = d(1, r_i(1)) + \sum_{j=2}^{n_i} d(r_i(j-1), r_i(j)) + d(r_i(n_i), 1)$$

and the total mileage covered will be on all routes $\sum_{i=1}^{k} D_i$.

The problem variables are the $r_i(j)$ values, the n_i values and k all of which must be positive integers and the $r_i(j)$ must be distinct. We wish to minimize k and, for minimum k, we wish to minimize $\sum_{i=1}^{k} D_i$, subject to the mileage and capacity restrictions that

$$D_i \leq M, i = 1, ..., k$$

$$\sum_{j=1}^{n_i} q(r_i(j)) \leq Q \text{ for } i = 1, ..., k.$$

(It is assumed that $Q \geq q(i)$.)

Several heuristic methods have been proposed for solving this problem to obtain near optimal solutions. In these heuristic methods the routes are built up link by link until all customers have been visited. There are various measures of priority for selecting a link. One such measure suggested by

Clarke and Wright is based on what is called the savings criterion. The saving $s(a, b)$ for the link between points a and b is measured as the mileage saved if a route previously ending at a is extended to include b rather than have a special route from the depot to include a and b on their own. Thus

$$s(a, b) = d(1, b) + d(1, a) - d(a, b).$$

At each stage in the heuristic procedure we consider all the possible links which have not yet been considered and measure the savings for each of these. The link with the maximum savings is selected. At the first iteration the highest priority link starts the first route (providing load and mileage restrictions are not broken). The second link is then considered and joined to the first if there is a point in common. All linkages involving the common point are deleted from the list. Otherwise the new linkage starts a second route. We then choose the next linkage which may possibly join two routes; if this happened it would mean that other linkages involving the two common points should be deleted from the list for further selection. At each iteration we must ensure that the mileage and capacity restrictions are satisfied. This means that we must not build up a route whose total mileage will exceed the maximum M nor must we allow a link to be included which would make a closed circuit. Thus at any stage a multiplicity of routes are being built up. The procedure is illustrated on the following example.

Example 7.5

Suppose we have nine points and a depot distributed in a plane with coordinates as shown in Fig. 7.3 with the depot being point number 1. First we

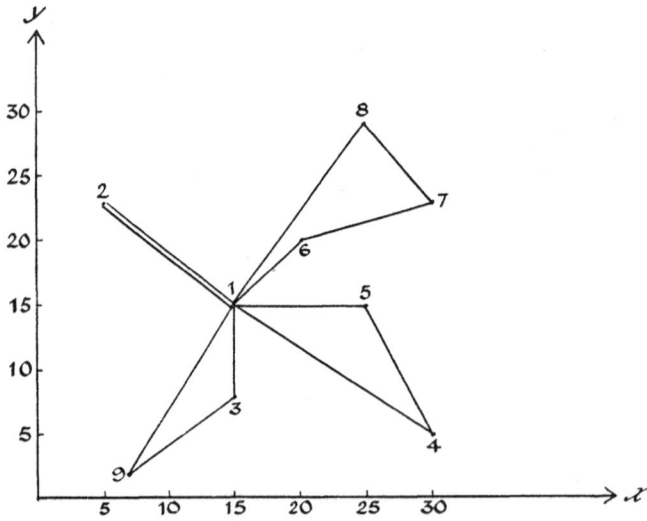

Fig. 7.3

note the distances of the points from the depot as shown in Table 7.12.

Table 7.12

Point number	2	3	4	5	6	7	8	9
Distance from depot	13	7	18	10	7	17	17	15

Then the distances and savings measures associated with the various possible links are as follows:

Table 7.13

Link	Distance	Savings measure	Link	Distance	Savings measure	Link	Distance	Savings measure
2–3	18	2	3–7	21	3	5–8	14	13
2–4	31	0	3–8	23	1	5–9	22	3
2–5	21	2	3–9	10	12	6–7	9	15
2–6	15	5	4–5	11	17	6–8	10	14
2–7	25	5	4–6	18	7	6–9	22	0
2–8	21	9	4–7	18	17	7–8	8	26
2–9	21	7	4–8	24	11	7–9	31	1
3–4	15	10	4–9	24	9	8–9	32	0
3–5	12	5	5–6	7	10			
3–6	13	1	5–7	10	17			

We set the mileage limitation as 45 and assume that there are no carrying limitations. The heuristic technique now determines the routes by the following decisions.

Step 1. Select the link 7–8 as the one with the maximum savings.

Step 2. Select link 4–5.

Step 3. It would be desirable to select link 4–7 but this would create a route which when joined to the depot would have a total distance in excess of 45 miles.

Step 4. Link 5–7 must similarly be rejected.

Step 5. Select link 6–7 and attach it to 8–7. Eliminate all links with point 7.

Step 6. Link 6–8 would be the next to be selected but this would close a circuit and it is therefore not feasible.

Step 7. Link 5–8 would be next, but it is not feasible.

Step 8. Link 3–9 is selected.

All remaining links are found to be non-feasible. Thus there are four routes as shown in Fig. 7.3.

Route 1: 1–6–7–8–1

Route 2: 1–5–4–1

Route 3: 1–3–9–1

Route 4: 1–2–1

and the total distance is $41 + 39 + 32 + 26 = 138$.

This method and other heuristic methods were applied by Gaskell to a number of large cases which included capacity restrictions. The results were compared with the best routes which could be obtained after a long and detailed visual inspection. The total mileages which were obtained for the visual and heuristic savings methods are recorded in Table 7·14. It may be

Table 7.14

Case	1	2	3	4	5	6
Mileage visual	851	1416	813	585	876	949
Mileage savings heuristic	923	1427	839	598	963	955

presumed that the visual solution is very close to the global optimum. If this is the case the heuristic method has usually demonstrated a reasonable performance and provides a quick systematic computation.

The vehicle scheduling problem may be generalized to more complex situations in which vehicles of different capacities have to be scheduled from a number of depots. So far few studies have been conducted of these more general contexts.

7.7 Timing problems: transport fleet scheduling and railway diagramming

Another group of vehicle scheduling problems arises when we have timing requirements on the vehicles. These timing restrictions may be enforced by demand considerations or there may be restrictions on the times at which the vehicles themselves can be moved. The first type of situation arises in the scheduling of buses to satisfy a given timetable or when a limited tanker fleet is available to meet a demand schedule at ports. In these cases we want to meet the schedule with the minimum number of vehicles. This requires an efficient organization of the vehicles so that once a vehicle has been used on one trip it can be employed on the next. It is equivalent to the maximization of vehicle utilization, and we will illustrate the basic idea of the scheme on an elementary tanker scheduling problem.

Example 7.6

Shiploads of a given capacity are to be sent from ports P_1 and P_2 to destinations Q_1, Q_2 and Q_3 on the following days of departure after a given start day, say day 0.

Table 7.15

Time of departure	0	8	32	49
From port	P_2	P_1	P_1	P_2
To destination	Q_3	Q_1	Q_2	Q_3

The travel times taken by loaded and unloaded ships are assumed to be the same in either direction and they are given in Table 7.16. The ships are

Table 7.16

		To		
		Q_1	Q_2	Q_3
From	P_1	21	19	13
	P_2	16	15	12

assumed to be all of the same type. The required schedule is to be satisfied using as few ships as possible. This is equivalent to maximizing the number of ships which can be used for second and third trips. We therefore need to consider the times at which ships reach their destinations and become free to make a new run from either of the ports. The arrival times are computed from Tables 7.15 and 7.16. For instance the ship leaving P_1 at time 32 for Q_2 will arrive at time $32+19 = 51$.

Table 7.17

From	P_2	P_1	P_1	P_2
To	Q_3	Q_1	Q_2	Q_3
At	12	29	51	61

We can now write out a list of required starting times and a list of destination arrival times using Tables 7.15 and 7.17 as shown in Table 7.18.

Table 7.18

Terminating schedule	1	2	3	4
Place	Q_3	Q_1	Q_2	Q_3
Time	12	29	51	61
Starting schedule	1	2	3	4
Place	P_2	P_1	P_1	P_2
Time	0	8	32	49

This table is inspected to see if some of the ships in the terminating schedule can be returned to P_1 or P_2 in time to do another journey in the starting schedule. The ships in the various positions in the terminating schedule are regarded as potential suppliers to the requirements of the starting schedule with restrictions on which transfers are possible. For example, the ship in the first position of the terminating schedule can be used to satisfy the third or fourth requirement in the starting schedule as this ship can reach P_1 at time $12+13 = 25$ or P_2 at time $12+12 = 24$. However, it can be seen at once that the third or fourth ships of the terminating schedule cannot be used as another starting ship. We therefore have a standard transportation problem with restrictions of the form mentioned in Section 7.4.

The problem variables are the possible transfers from the terminating schedule to the starting schedule. Let x_{ij} be one or zero according to whether a ship is transferred from terminating position j to starting position i. Then we wish to maximize such transfers, i.e.

$$\text{maximize } \sum_{i=1}^{4} \sum_{j=1}^{4} x_{ij}.$$

The constraints are that $x_{ij} = 0$ where it is not possible to make the transfer. In this example $x_{ij} = 0$ for all i and j except x_{31}, x_{41} and x_{42}. Also there are constraints to ensure that not more ships are sent back than necessary, i.e.

$$\sum_{j=1}^{4} x_{ij} \leqq 1 \text{ for all } i.$$

Finally, all $x_{ij} \geqq 0$.

The solution to this particular problem is to set $x_{31} = 1$ and $x_{42} = 1$. The formulation can readily be generalized to the situation where more than one ship is despatched at each starting position and more than one ship is available. This case is studied as an exercise. The ship scheduling problem has also been formulated in a more general framework by Briskin where the days of departure are themselves variables, and dynamic programming has been used to solve this case.

Timing considerations may also arise on vehicle movements where there are restrictions on the times at which the vehicles themselves can be moved. The most important of these problems arises in railway systems where there are limitations on the density of traffic flow along a line and the timing of one train depends on the timing of predecessor trains on the same track, there being only a few points at which it may be possible to overtake.

The problem of scheduling a sequence of trains of differing speeds down a stretch of track can be illustrated graphically. Fig. 7.4 shows 5 trains numbered 1 to 5 travelling from A to D and passing through two potential overtaking points B and C. The horizontal axis shows time increasing and the vertical axis is distance along the line from A to D. The relative slopes of the lines indicate the speeds of the individual trains on the different stretches. Train numbers 2 and 4 stop at stations B and C and trains 1, 3 and 5 run non-stop. The timing of all the trains over the stretch of track are clearly interdependent. The planning of timetables by this type of graphical procedure is known as railway diagramming.

Two optimization problems arise in railway diagramming. The first concerns the design of a basic timetable. Here we wish to consider the demand at the various points over time, but at the same time generate a reliable timetable. A reliable timetable is one in which there is not too much interaction between the timings of trains, enabling a schedule to be maintained despite individual deviations from the planned timings. The second form of optimization problem is to find the best paths through a partially planned timetable for running of extra goods traffic or returning locomotives to start points. The best path will be the path amongst a number of possibilities which minimizes

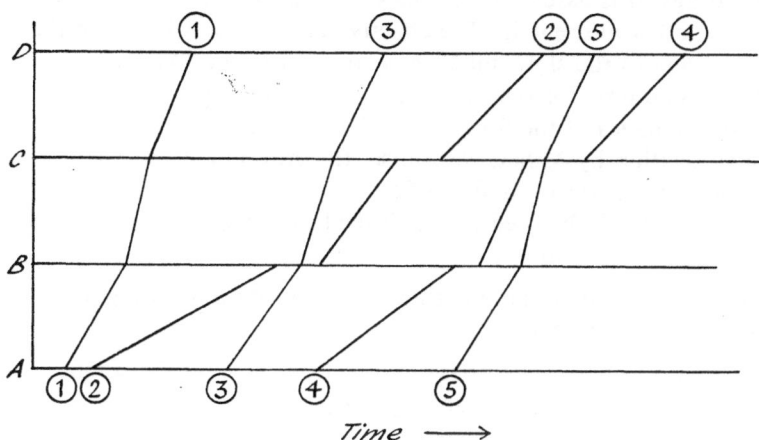

Fig. 7.4

the time of travel; the travel time includes lost time spent in loops and sidings to avoid disturbing the existing timetable. Both these problems are difficult to express mathematically and there are serious implementation problems. However, an efficient automated technique for railway diagramming might save a great deal of clerical effort and offer improvements in control. At present no general formulation of these problems seems to have been made. However there is no doubt that considerable skills have been accumulated by experienced staff on railway diagramming problems, and their techniques must embody efficient, if unpublished, heuristics.

*7.8 Traffic signal setting

In this final section two optimization problems arising in the setting of traffic light signals will be reviewed. The first problem is to determine the optimal settings at a two-way intersection where traffic rates are different in different directions and are subject to statistical variation. The second problem concerns the best synchronization of a series of traffic lights down a length of road or motorway containing a number of crossings.

A fair proportion of traffic signal settings are operated dynamically by adjusting the settings to the traffic flow. The lengths of the green and red phases vary from cycle to cycle according to demand, and the pneumatic pads on each approach transmit the signals. However, in heavy traffic the

G

queue lengths are inadequately communicated through the pads and there is a need to know how to fix the signal settings. The problem is examined in the next example.

Example 7.7

Suppose we have a simple four-way junction for which we wish to determine the fixed signal settings. It is assumed that the signals go green in one direction and then in the other alternately. The effective green time is less than the actual green time owing to the slow initial movement of traffic. We wish to determine the total cycle time for the signals and its division into the two directions given the statistics of the flow of traffic in the four directions. Following the investigation by Miller, we will set the criterion as the minimization of the total rate of delay to all traffic.

The following notation will be used in which the four directions are assumed to be numbered 1, 2, 3, 4 in rotation.

c = the cycle time of the system, i.e. the time from a given direction going green until it goes green again.

g_i = the length of the effective green phase in direction i, (as the directions are numbered in rotation $g_1 = g_3$ and $g_2 = g_4$).

q_i = the average rate of arrival of vehicles from direction i.

s_i = the 'saturation flow' in direction i, i.e. the average departure rate during the effective green phase while the queue remains.

L = total lost time per cycle (i.e. $L = c - 2(g_1 - g_2)$).

r_i = the ratio of the variance of arrivals to the mean count of arrivals from direction i.

$p_i = g_i/(c-L)$ the proportion of the total effective green time allocated to direction i.

The problem variables are the total cycle time i and the p_i values. (This effectively includes the g_i values as the p_i values depend on the g_i quantities and c.)

The objective function is the total rate of delay to all traffic for any given values of the problem variables. This measure depends on all the quantities in the above list and Miller derives a formula for it by making some reasonable assumptions and using queueing theory. We will not reproduce the derivation here but merely quote the formulae. The total rate of delay is

$F(c, p_1, p_2, p_3, p_4)$

$$= \sum_{i=1}^{4} \frac{q_i(1 - p_i(1 - L/c))}{s_i - q_i} \left\{ \frac{s_i r_i}{q_i} \left(\frac{q_i c}{s_i p_i(c-L) - q_i c} - 1 \right) + s_i(c - p_i(c-L)) \right\}.$$

The only constraints on the problem variables are that they should all be positive.

We now wish to minimize F with respect to c and the p_i variables, and we can do this by a non-linear optimization technique. It can also be achieved by calculating the derivatives $\dfrac{\partial F}{\partial c}$ and $\dfrac{\partial F}{\partial p_i}$ and making some approximations to determine explicit values for the minima from the derivative equations.

It is interesting to note the forms of the functions for the rate of delay for different flow rates. They are illustrated in Fig. 7.5. In all cases the delay

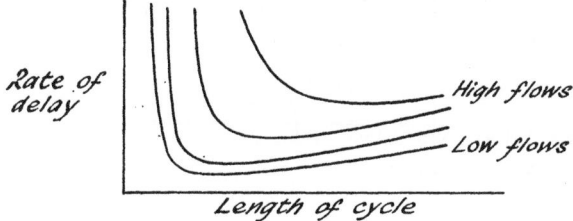

Fig. 7.5

increases rapidly for cycle lengths less than the optimum, but only slowly if the cycle length is too long. Also, to err on the safe side, high flow rates require longer cycle times than low flow rates. Therefore, optimum settings should be found for the direction of traffic with the highest traffic intensity.

The optimization procedure has been applied to real data. Suppose the average arrival rate from North, South, East and West were 1080, 720, 360 and 180 vehicles per hour as shown in Fig. 7.6. Also let the ratios of variance

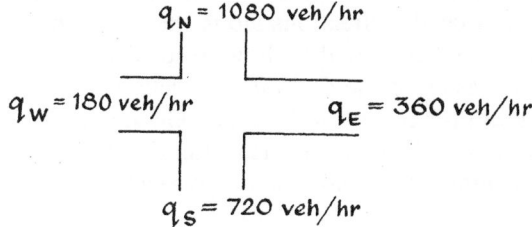

Fig. 7.6

to mean be $r_N = 1 \cdot 5$, $r_W = r_S = r_E = 1$. The traffic from the North therefore tends to be bunched and is random in the other directions. Assume the lost time per cycle is 12 seconds and that the saturation flow for all directions is 1800 vehicles per hour. For this data the optimum cycle time to minimize the rate of delay is 122 seconds and the proportion of the time to set the signals for the North-South direction is 0·75. These results are illustrated in the graph of Fig. 7.7 in which the rate of delay is plotted against cycle time for three possible p_i values and the p_i value for North-South direction is marked as $\bar{p}$.

The problem of synchronizing a series of traffic lights along a stretch of road in an optimal manner is more difficult. However, some optimization

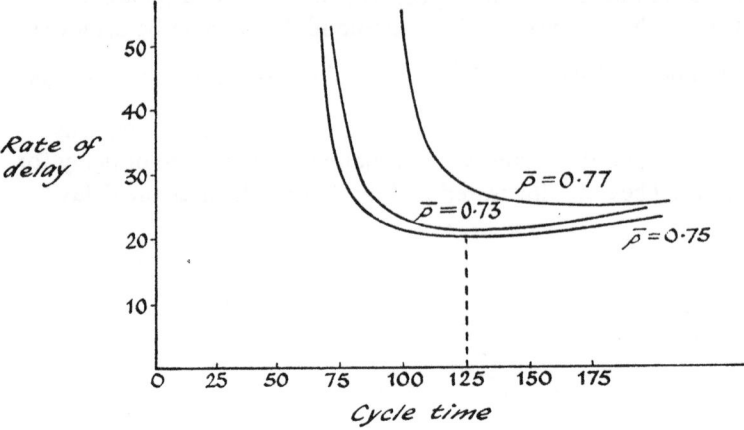

Fig. 7.7

studies have been made for these cases and we will review a linear programming formulation of one such problem investigated by Little.

Example 7.8

Traffic signals are to be placed along a two-way stretch of road at each of a number of cross-roads spaced along it. As the given stretch of road is a trunk road, the traffic engineers problem is to try to synchronize the signal settings so that if a driver maintains the correct speeds, he can go from one end of the stretch of road to the other without stopping. This is called the arterial problem. Suppose there are n signals and that the directions along the road are distinguished as outbound and inbound. The signals will be denoted by $S_1, S_2, ..., S_n$ with the subscript increasing in the outbound direction.

Fig. 7.8 shows a space-time diagram for travel on the road. The horizontal axis denotes time and the vertical axis denotes distance from the start of the given stretch of road. The horizontal lines inside the diagram are barriers

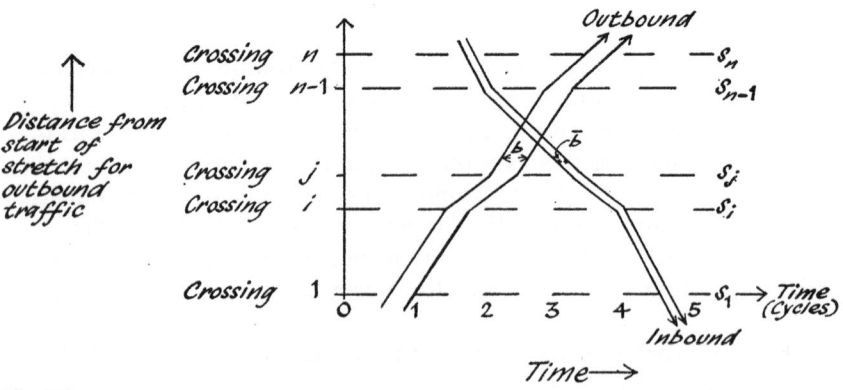

Fig. 7.8

to the traffic along the road indicating when the signals are red at the various crossings. The zigzag lines represent trajectories of the cars passing unimpeded along the road in the directions marked and changes in slope correspond to changes in speed. The set of possible unimpeded trajectories in a given direction forms the 'green band' whose horizontal width is the bandwidth for that direction for a given speed pattern. b is the bandwidth in the outbound direction and $\bar{b}$ is the bandwidth in the inbound direction. Clearly with fixed signal settings on the crossings and known common speed patterns for the vehicles the green bands occur in a parallel series across the diagram. Also as the traffic signals will be synchronized we can measure time in terms of the cycle of synchronization as marked on the diagram.

The general problem is to synchronize the signals so as to maximize some weighted function of the bandwidths thus getting the maximum amount of traffic through the system as smoothly as possible. Here we will study the simpler problem of determining how to synchronize the signals to produce the largest equal bandwidths in both directions. We will assume that the signals have a common period, i.e. the sum of the red and green times are equal over all signals.

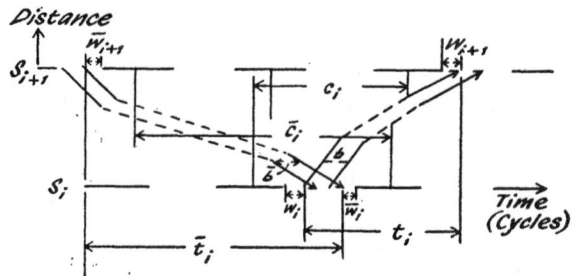

Fig. 7.9

This problem can be formulated as a linear programming problem. We define the following notation related to Fig. 7.9 for the traffic from signal S_i to S_i+1.

b = bandwidth in outbound direction.

r_i = red time in cycles at crossing i.

t_i = travel time in cycles from S_i to S_{i+1} in outbound direction.

c_i = time in cycles from centre of a particular red at S_i to centre of particular red at S_{i+1}. The two reds are chosen so that each is immediately to the left of the same outbound green band as shown in Fig. 7.9. c_i is measured positively if the centre at S_{i+1} lies to the right of the centre at S_i as it does for the outbound traffic in Fig. 7.9.

w_i = time in cycles from the right side of S_i's red to the green band.

The corresponding quantities for the inbound direction from signal (S_{i+1}) to S_i are denoted by the same letters and suffices with bars, $\bar{t}_i$, $\bar{c}_i$, $\bar{w}_i$.

The problem variables are the bandwidth b and the (off-setting) of the signal times in relation to the green band w_i and $\bar{w}_i$. The objective is to maximize the band width, i.e.

maximize b.

The first constraints require that the band size b should fit within the green period. From Fig. 7.9 we require that

$$w_i + b \leq 1 - r_i, \quad i = 1, \ldots, n$$

and

$$\bar{w}_i + b \leq 1 - r_i, \quad i = 1, \ldots, n.$$

The second set of constraints are more complex. We require that for the particular outbound and inbound traffic streams which go through the same green phase at crossing i there must be an integral number of cycles between the green phases which they pass through at crossing $(i+1)$. This situation is illustrated in Fig. 7.9. In effect we require that $(c_i + \bar{c}_i)$ must be an integer. From the diagram

$$c_i = \tfrac{1}{2}r_i + w_i + t_i - w_{i+1} - \tfrac{1}{2}r_{i+1}$$

and

$$\bar{c}_i = \tfrac{1}{2}r_i + \bar{w}_i + \bar{t}_i - \bar{w}_{i+1} - \tfrac{1}{2}r_{i+1}.$$

Adding these up, and denoting any positive integer by m_i, we require that the following two constraints should hold:

$$w_i + \bar{w}_i - (w_{i+1} + \bar{w}_{i+1}) + (t_i + \bar{t}_i) = m_i - (r_i - r_{i+1})$$

where m_i is integral valued.

Finally the variables b, w_i and $\bar{w}_i$ must all be non-negative. This completes the linear programming formulation of the arterial traffic signalling problem.

Little goes on to generalize this problem to include the traffic speeds and the signal setting periods as decision variables, and demonstrates how the branch and bound method for integer linear programming may be used to solve the problem optimally for a moderately large number of variables.

REFERENCES

Balinski, M. L. and Quandt, R. E. 1964. On an integer programme for a delivery problem. *Opns Res.*, **12**, 300.

Briskin, L. E. 1966. Selecting delivery dates in the tanker scheduling problem. *Man. Sci.*, **12** (6), B 224.

Clarke, G. and Wright, J. W. 1964. Scheduling of vehicles from a central depot to a number of delivery points. *Opns Res.*, **12**, 568.

Dantzig, G. B. 1963. *Linear Programming and Extensions.* Chapter 17. Princeton University Press.

Dantzig, G. B. and Fulkerson, D. R. 1954. Minimizing the number of tankers to meet a fixed schedule. *Nav. Res. Log. Quat.*, **1**, 217.

Ferguson, A. R. and Dantzig, G. B. 1956. Allocation of aircraft to routes—an example of linear programming under uncertain demand. *Man. Sci.*, **3**, 45.

Ford, R. L. and Fulkerson, D. R. 1957. A simple algorithm for finding maximal network flows and an application to the Hitchcock problem. *Can. J. Math.*, **9**, 210.

Frank, O. 1966. Two-way traffic on a single line of railway. *Opns Res.*, **14**, 801.

Gaskell, T. J. 1967. Bases for vehicle fleet scheduling. *Opns Res. Quat.*, **18**, 281.

Lee, A. M. 1964. Transportation. Chap. 8 in *Progress in Operations Research*. Vol. II, Ed. D. B. Hertz and R. T. Eddison. Wiley, New York.

Little, J. D. C., 1966. The synchronization of traffic signals by mixed integer-linear programming. *Opns Res.*, **14**, 568.

Miller, A. J. 1963. Settings for fixed cycle traffic signals. *Opl. Res. Quat.*, **14**, 373.

Nicholson, T. A. J. 1966. Finding the shortest route between two points in a network. *Comp. J.*, **9**, 275.

Vajda, S. 1958. *Readings in Mathematical Programming*. Pitman, London.

Exercises on Chapter 7.

1 A bus company needs to despatch relief buses to points P_1, P_2, P_3, P_4. The quantities required are 1, 6, 2 and 6 respectively and there are 5 spare buses at each of the garages G_1, G_2, G_3. The time in minutes taken to travel from each garage to each destination are given in Table 7.19.

Table 7.19

		To		
	P_1	P_2	P_3	P_4
G_1	5	4	3	2
From G_2	10	8	4	7
G_3	9	9	8	4

The manager would like to distribute the buses in such a way that the total amount of time taken from the garages to the destinations is as small as possible. Formulate the corresponding optimization problem, and solve it by the special transportation method described in Section 7.3.

2 Suppose that in Example 7.1 it has been possible to transport the coal to the power stations via other sources or destinations, and that the transportation costs between sources and between destinations is as given in the following tables.

Table 7.20

		To	
	S_1	S_2	S_3
S_1	0	3	2
From S_2	2	0	3
S_3	2	8	0

Table 7.21

| | To | | | |
	D_1	D_2	D_3	D_4
D_1	0	5	2	3
D_2	5	0	1	8
D_3	4	2	0	4
D_4	5	3	9	0

Express the transhipment problem of routing the coal from sources to destinations in the most economic way assuming symmetrical distances from sources to destinations.

3 Apply the dynamic programming calculation to find the shortest route between points 1 and 8 in the network in figure 7.10 where the distances are marked on the links.

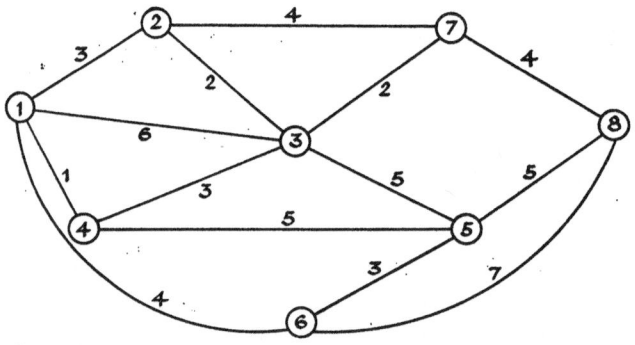

Fig. 7.10

4 In Example 7.5 on vehicle scheduling the mileage limit was 45 and there were no capacity limitations. Suppose the mileage limit was increased to 70 and the capacity limit was 10 with loads at demand points as given in Table 7.22.

Table 7.22

Point	2	3	4	5	6	7	8	9
Load	3	4	3	2	5	4	1	4

Using the data on distances between the 9 points as prepared for Example 7.5, what routes will now be determined by the heuristic procedure?

5 A firm operating a transport fleet wishes to know how many vehicles to own itself and how many to hire. (The vehicles are all assumed to be of a standard type.) The demand for transport varies from day to day over N days and the demand on day i is d_i (measured in vehicle loads). The capital cost of owning a vehicle is c and the variable cost of running it for a day is v. The cost of hiring a vehicle of the same type is h. If the demand must be met, formulate the problem of determining how many vehicles the firm should own to minimize total costs over the period.

6 A ship fleet operator wishes to plan his fleet schedule so as to maximize ship utilization. The current demands on the fleet are given in Table 18.23 which shows the times and numbers of ships which are required to go from the ports P_i to the destinations Q_j. The number of ships are given in parentheses.

Table 7.23

		To	
	Q_1	Q_2	Q_3
P_1	8(2)	14(1)	35(1)
From P_2	16(2)		46(1)
P_3		0(1) and 32(1)	

The sailing times between the ports and destinations (the same in either direction) are:

Table 7.24

	Q_1	Q_2	Q_3
P_1	10	4	6
P_2	5	5	8
P_3	6	11	3

Formulate the problem of maximizing the utilization of the fleet in integer linear programming terms.

7 Railway trains have to be sequenced through a junction and it is sometimes necessary to delay a train on one line to let another train through. If a train on route i passes through the junction (the route distinguishing an incoming as well as an outgoing direction) a minimum time of $t(i, j)$ minutes must elapse before another train on route j may pass through the junction. If trains have arrival times a_i on route i (we assume at most one train on one route at one time) formulate the optimization problem of determining the order in which the trains should be sequenced through the junction to mini-mize the maximum delay to any train.

*8 Formulate the vehicle routing problem discussed in Section 7.6 as a permutation problem.

*9 Confirm that the stopping rule of the dynamic programming calculation of the shortest route between two points in a network is in fact valid.

8 Planning in investment and advertising

8.1 Planning under uncertainty

Most of the previous applications have been allied to the production control activities within industry. In this final chapter two planning aspects will be studied which are less production oriented but are nevertheless fields in which optimization techniques are beginning to be used. It may seem unusual to combine finance and advertising in a single chapter but they share two common themes. Both are concerned with the efficient use of funds over time and both are largely dominated by uncertainties of the future. The speculator places his funds so as to maximize his expected but uncertain returns and the advertisements are placed in media to stimulate an unknown but hoped-for market response.

We will investigate four cases of the use of optimization techniques in financial investment. The first case is static and examines the problem of selecting a portfolio of securities for a given quantity of cash subject to limitations on the allowed risk. The second case takes account of timing considerations for the refunding of bonds: the problem has a very similar nature to the machine replacement studies solved by dynamic programming in Chapter 6. The third case examines how to balance the choices which may confront a firm for financing its seasonal cash requirements, and the subsequent case investigates how a firm may allocate its funds across interdependent projects over time so as to satisfy certain restrictions on earnings. Both these problems are formulated in linear programming terms.

The application of optimization techniques to problems in advertising is rather less developed than the financial applications. However, it is important to recognize how optimization problems arise in this field if data of adequate reliability can be obtained. The major difficulty is to construct effective models of consumer reaction to advertising. We will make some simple assumptions and examine the problem of deciding which media to insert advertisements in, and, in the dynamic context, how to maintain a given level of awareness about a product over a specified period at minimum cost.

8.2 Portfolio selection

The problem of selecting the best securities for the investment of a fund is not difficult to formulate if it can be assumed that we have reliable statistical information on which to assess future returns and the associated risks. Suppose a fund of £C is available for investment and a total of N securities are considered. Let the expected value and the variance on the return from security i be μ_i and σ_{ii}. The variation between securities may be correlated, and σ_{ij} denotes the convariance between the returns from securities i and j. If the price of security i is p_i we need to determine how much of each security to buy to achieve a given objective. Let x_i be the quantity of each security which is bought. Then if the objective is to maximize the expected return, we obtain the objective function

$$\text{maximize } \sum_{i=1}^{N} \mu_i x_i$$

$$\text{subject to } \sum_{i=1}^{N} x_i p_i \leqq C$$

$$x_i \geqq 0.$$

Clearly this problem is very simple to solve as we will invest in the security with the largest return. However we may also want to ensure that the variation in the return is not too large. For example we may insist that the variance be less than a certain quantity V. In this case we will need the additional constraint:

$$\sum_{i=1}^{N} \sum_{j=1}^{N} \sigma_{ij} x_i x_j \leqq V.$$

The problem is now not so simple, but it can still be solved by a straight forward non-linear optimization technique.

The risk factor may so dominate the investment policy that the objective and constraint may be turned around giving a problem of obtaining a specified return at a minimum possible variance. In this case the last constraint and the objective function are interchanged and we minimize the variance

$$V = \sum_{i=1}^{N} \sum_{j=1}^{N} \sigma_{ij} x_i x_j$$

subject to obtaining a minimum specified level of return M

$$\sum_{i=1}^{N} \mu_i x_i \geqq M$$
$$x_i \geqq 0$$
$$\sum_{i=1}^{N} x_i p_i \leqq C.$$

The non-linear part of the problem is now contained in the objective function.

Although it is possible to estimate the quantities μ_i and σ_{ij} from back records of stock market data (and this has indeed been done—see the reference by Sharpe), we are still making the assumption that security returns can be characterized by a mean value and a variance. It would be more accurate to consider how these parameters themselves change over time. But this would introduce a further depth of uncertainty which we will not investigate here.

The problem of buying and selling on the stock market (rather than choosing securities for a permanent investment) is characterized by the same difficulties of uncertainty. Although, theoretically, an optimal policy for stock market dealings would be to sell all shares when their prices are at a local maximum and buy at a local minimum as shown in Fig. 8.1, it is impossible

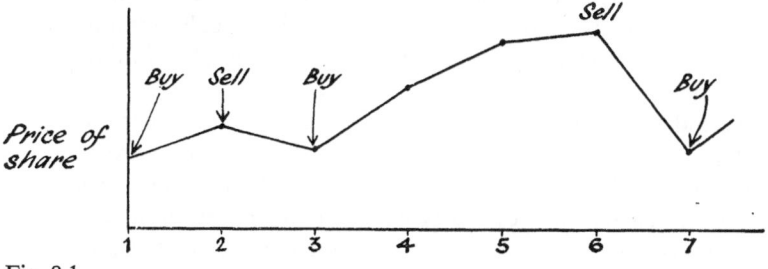

Fig. 8.1

to organize this in practice because of the unpredictability of share prices.

*8.3 Optimal timing of bond refunding

In some areas of financial control uncertainty about the future is not so dominant as to prevent a useful analysis of the decisions being undertaken. There are a number of financial holdings such as bonds and loans which have planned durations and which may be cut short part way through their duration at a certain cost if current conditions suggest this is desirable. We will review one particular study in this area by Weingartner for optimizing the timing of bond renewals.

When bonds are issued they have a specified lifetime or period to maturity, and a given coupon rate or yield. However, bonds frequently contain a provision whereby the issuer can call or redeem the bonds before they have matured. A certain amount of notice has to be given for this to be done and a payment of a premium has to be made. From the issuers' point of view, the bond is redeemed in order to reduce the interest cost on borrowed funds, as it may turn out to be possible to issue new bonds at lower interest rates. The decision is therefore referred to as bond refunding.

Basically the refunding decision must weigh the alternatives of continuing payment of the interest on the outstanding bonds, or making a present payment of the cost of floating a new issue plus the call premium in order to achieve lower future interest rates on the new issue. It is assumed therefore that future interest rates and bond yields are known, and that the various cost factors can be measured. In making the refunding decision the stream of future payments must be discounted.

The refunding problem is very similar to the plant renewal study discussed in Section 6.3 and the same dynamic programming treatment can be applied. We will assume for simplicity of notation that the refunding decision is made once per interest period (for instance once a year) and that all costs are expressed in terms of £ per £ of existing debt so that the size of the bond issue can be kept out of the problem. All interest rates will be measured per interest period. Calendar time is denoted by $t = 0, 1, 2, ..., T$, where T is the horizon, and we assume that the bond is of age j_0 at time $t = 0$. Also let $r_t(m)$ denote the interest payed (coupon rate) for a bond with an original maturity of m periods after issue in period t. In general, the call premium will be a declining function of age and original term to maturity. (The initial call premium may be equal to one year's coupon.) The call premium for a bond issued in period t and of age j with a term to maturity of m periods is denoted by $c_j(r_{t-j}(m), m)$. The fixed costs of floating a new issue are denoted by F. Finally we denote the rate of interest for discounting purposes in year t by d_{t-j} if the current age of the bond is j. (This notation takes account of the effect the refunding policy may have on the estimated discounting rate.)

The objective is to plan the refunding operations so as to minimize the total discounted costs. The problem variables are the points in time in the interval $0 \leq t \leq T$ at which refunding should take place. It is assumed for simplicity that bonds which are issued for a refunding operation will have a maturity m fixed by convention. However, bonds issued near the end of the process will be assumed to mature at the horizon T.

In order to use dynamic programming we define the following optimality function $f_t(j)$ for the recurrence relations:

$f_t(j) =$ the discounted cost associated with the liability of bonds of age j in year t when an optimal refunding policy is followed for the remainder of the process.

In dynamic programming terms the period t represents the stage and the age of the bond j represents the state. At period t there are two possible decisions: either to keep the bond for a further period or to refund it. The cost of keeping is the coupon cost in this period $r_{t-j}(m)$ plus the least future costs from period $(t+1)$ onwards starting with a bond of age $(j+1)$, all discounted at a rate of interest d_{t-j}; i.e. the keeping cost is

$$\frac{r_{t-j}(m)+f_{t+1}(j+1)}{1+d_{t-j}}.$$

The cost of refunding is the cost of the call premium $c_j(r_{t-j}(m),\ m)$, the cost of floating a new issue F, together with the coupon cost of a bond issued in this period $r_t(m)$ plus future costs with a bond of age 1 discounted at the new rate of interest d_t, i.e. the refunding cost is

$$c_j(r_{t-j}(m),\ m)+F+\frac{r_t(m)+f_{t+1}(1)}{1+d_t}.$$

Therefore from state j at stage t we can move into state $(j+1)$ or state 1 at stage $(t+1)$ and the dynamic programming recurrence relation is

$$f_t(j)+\min \begin{cases} \text{Keep:} & \{r_{t-j}(m)+f_{t+1}(j+1)\}/(1+d_{t-j}) \\ \text{Refund:} & c_j(r_{t-j}(m),\ m)+F+\{r_t(m)+f_{t+1}(1)\}/(1+d_t). \end{cases}$$

As the period ends in year T we can write $f_{T+1}(j) = 0$ for all j. The formula must be modified when the period t is within m periods of the horizon T to ensure that the bond matures at the horizon T. This is achieved by substituting

$$\min\ (T-t+j,\ m)$$

for m where m occurs in $r_{t-j}(m)$ and $c_j(r_{t-j}(m),\ m)$. The formulae are then evaluated for $t = T,\ T-1,\ ...,\ 2,\ 1$, and for values of j in the range $1 \leq j \leq m$.

Weingartner extends this dynamic programming model of the refunding problem in situations where the bonds may have a variable maturity and the horizon may be infinite. He also considers various aspects of uncertainty and points out how the 'backwards calculations' of dynamic programming (i.e. working from period T to period 1) can accommodate a flexible and subtle assessment of the discounting factor d_{t-j} where it is dependent on the initially unknown refunding policy.

8.4 Financing seasonal cash requirements

Many companies are faced with the task of choosing a source of short-term funds from a set of financing alternatives. If one of the alternatives is clearly superior to all the other alternatives, the short-term financing problem is easily resolved by accepting that alternative. Sometimes, for example, the company may be able to obtain all the funds it needs from a favourable line of credit. More frequently, the situation is that either one alternative is not superior in every respect over the others or there are various constraints placed across the alternatives. In these circumstances some of a number of alternatives may be selected as the package solution.

Example 8.1

We will consider a typical short-term financing problem of a company subject to seasonal cash flow patterns. Suppose that a company has the following cash requirements over 10 periods. The requirements are measured in 1000£.

Table 8.1

Period	1	2	3	4	5	6	7	8	9	10
Cash requirement	1000	1500	2000	−400	−500	−4500	400	1200	3000	−500

The negative signs show where cash surpluses will occur. The requirements may be graphed as shown in Fig. 8.2 The cash requirements are calculated by subtracting the total receipts in each period from the total disbursements. The receipts consist of accounts to be received and the disbursements consist of accounts to be payed, payments for purchases, etc.

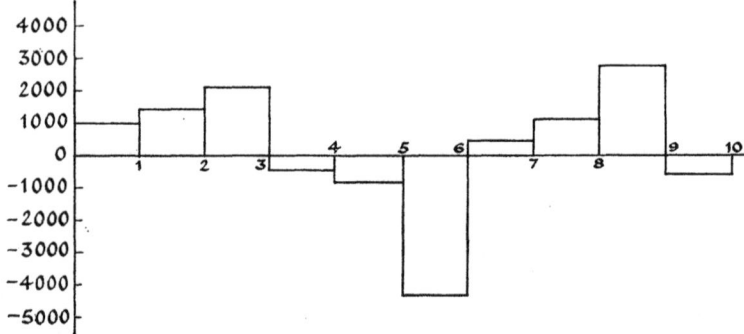

Fig. 8.2

There may be a number of alternative ways of financing the cash requirements each with its own costs and conditions. We will consider the following set of possibilities.

(i) Unsecured line of credit

The company can borrow up to £1,500,000 from a bank without having a special security. The interest rate in this line of credit is 0·9 per cent per period.

(ii) Bank loan using the security of accounts receivable

A bank will lend up to 80 per cent of the value of the accounts receivable. The cost of borrowing under this alternative is 1·4 per cent per period of the average amount of the loan outstanding during the period. The bank will not permit a loan to be taken out in combination with an unsecured line of credit.

(*iii*) Delaying of accounts payable

The company may delay payments of accounts payable. If such payments are delayed, a discount is lost which on the average amounts to $3\frac{1}{2}$ per cent. The payments may not be delayed for more than one period.

(*iv*) Term loan

The company has an offer of a term loan from its bank. The term loan must be taken out at the beginning of the initial period; it is limited to a maximum amount of £2,000,000 and it cannot be taken out in an amount less than £750,000. The principal of the loan must be repaid in eight equal installments, the first installment is due six periods after the loan is initially taken out, with the subsequent installments due at six-period intervals. The interest rate on the term loan is 0·8 per cent per period. If the firm takes out a term loan, the bank insists on further constraints on the combined borrowing under the term loan and either the line of credit or the secured loan. The maximum borrowing under the term loan plus a line of credit is £2,500,000, while the maximum borrowing with a term loan and a secured loan is £4,000,000.

The problem facing the company is to devise the cheapest means of financing its cash requirements. The immediate decision is to choose whether to take out a term loan, and, if so, what amount. Furthermore, it has to be decided how to acquire additional cash for each period if it is needed or what to do with excess cash if any is available. We will not consider the problem of dealing with excess cash, and set up the objective of minimizing the total cost.

The problem variables are the quantities of the various sources of funds which should be taken on in each period. Let us number the four sources:

Source 1: unsecured line of credit

Source 2: bank loan with security of accounts payable

Source 3: delaying of accounts payable

Source 4: term loan.

We have defined the following variables over 10 periods:

x_{ij} is the amount borrowed from source i in period j

y_{ij} is the amount voluntarily repaid to source i in period j

v_{ij} is the amount of compulsory repayment to source i in period j

z_{ij} is the net cumulative amount borrowed from source i in period j after borrowing and repayment for period j.

Also the following data is assumed to be known:

r_i is the rate of interest for borrowing from source i for one period

a_j is the amount of accounts receivable in period j

p_j is the amount payable in period j due to purchases.

The objective function to be minimized consists of the interest payments on loans and discount costs.

The cost in period j is

$$r_1 z_{1j} + r_2 z_{2j} + r_3 z_{3j} + r_4 z_{4j}$$
$$= 0{\cdot}9 z_{1j} + 1{\cdot}4 z_{2j} + 3{\cdot}5 z_{3j} + 0{\cdot}8 z_{4j}.$$

Therefore the total costs are

$$\sum_{j=1}^{10} \left(0{\cdot}9 z_{1j} + 1{\cdot}4 z_{2j} + 3{\cdot}5 z_{3j} + 0{\cdot}8 z_{4j} \right).$$

The first constraints are the equations relating the z_{ij} quantities to x_{ij}, y_{ij}, v_{ij}. The net amount borrowed clearly equals the sum of the borrowings less the repayments. These relationships may be expressed as

$$z_{ij} = \sum_{k=1}^{j} (x_{ik} - y_{ik} - v_{ik}) \text{ for } i = 1, 4; \ j = 1, 10.$$

All the quantities x_{ij}, y_{ij}, v_{ij}, z_{ij}, must be non-negative.

The remaining constraints express the limitations and conditions on the loans from the separate sources. On the unsecured line of credit the borrowing is restricted to a limit of £1,500,000. We therefore require

$$\sum_{k=1}^{j} (x_{1k} - y_{1k}) \leq 1{,}500{,}000 \text{ for } j = 1, 2, \ldots, 10.$$

Also the voluntary repayments must not exceed the amount borrowed. Therefore

$$\sum_{k=1}^{j} (x_{1k} - y_{1k}) \geq 0 \text{ for } j = 1, 2, \ldots, 10.$$

The size of the secured bank loan is limited to 80 per cent of the value of the accounts receivable. As a_j is the amount of accounts receivable then

$$\sum_{k=1}^{j} (x_{2k} - y_{2k}) \leq 0{\cdot}8 \, a_j.$$

As the bank will not allow both a line of credit and a secured loan simultaneously, we require that

$$x_{1j} = 0 \text{ or } x_{2j} = 0, \text{ for } j = 1, 2, \ldots, 10.$$

(Notice that this is a conditional constraint requiring 10 zero-one variables if it is to be converted into standard integer-linear form.)

Again repayments cannot exceed the amount borrowed.

$$\sum_{k=1}^{j} (x_{2k}-y_{2k}) \geqq 0 \text{ for } j = 1, 2, ..., 10.$$

The delaying of accounts payable is limited to one period. As p_j is the amount payable in period j due to purchases, the amount repaid is y_{3j} whereas the amount not paid and thus providing funds for the period is x_{3j}. As an account is either paid or not paid

$$x_{3j}+y_{3j} = p_j.$$

Then the amount x_{3j} must be compulsorily repaid in period $(j+1)$, i.e.

$$v_{3\ j+1} = x_{3j}.$$

Finally, the term loan can only be taken out in the first period, and is limited to a maximum of £2,000,000 and must be at least £750,000. Therefore either

$$750,000 \leqq x_{41} \leqq 2,000,000$$

or

$x_{41} = 0$. (This provides one further conditional constraint.)

Also

$$x_{4j} = 0 \text{ for } j = 2, 3, ..., 10.$$

The principal of the loan must be paid back in eight equal installments due at six-period intervals extending well beyond the 10 periods of the study Therefore

$$v_{4j} = \tfrac{1}{8}x_{41} \text{ for } j = 6k+1; k = 1, 2, ..., 8.$$

The outstanding balance is given by

$$z_{4j} = x_{41} - \sum_{k=1}^{j} v_{4k}.$$

The maximum borrowing under the term loan plus a line of credit is £2,500,000, while the maximum borrowing with a term loan and a secured loan is £4,000,000. Therefore

$$z_{4j}+ \sum_{k=1}^{j} (x_{1k}-y_{1k}) \leqq 2,500,000$$

$$z_{4j}+ \sum_{k=1}^{j} (x_{2k}-y_{2k}) \leqq 4,000,000.$$

This completes the definition of the optimization problem. All the functions are linear apart from the conditional constraints. Each of the possibilities associated with the conditional constraints can be accommodated through the introduction of eleven zero-one variables in the manner described in Chapter 11, and then the branch and bound method could be employed in conjunction with linear programming to obtain the solution directly. This problem has

been studied in the reference by Robichek *et al.*, where additional consideration is given to the uncertainty in the data and further qualitative factors are
considered in the objective function.

8.5 Allocation of funds to interdependent projects

Besides wishing to finance its cash requirements over time in an optimal way
a firm will also want to allocate its funds efficiently amongst alternative
investment projects. Several studies have been made of this type of problem,
most of which can be formulated in linear programming terms.

First we describe the most elementary model. It is assumed that there are
a set of N independent investment alternatives. The firm has to decide which
of these investment projects to invest in. Each alternative will yield cash flows
in a number of periods and a_{jt} is the estimated cash flow from project j in
period t. (There are a total of T periods.) If in period t there is a net outflow of
cash, a_{jt} is a negative quantity. The expenditures on capital account for project j in period t are denoted by c_{jt} and C_t is the ceiling imposed on capital
expenditures in period t. The objective is to maximize the net present value
from all projects which are accepted where r is the steady rate of discount
over the periods.

The problem variables reflect the decisions of whether or not to invest in
a project. Let x_j be unity if project j is accepted and 0 if it is not accepted.
Then the objective function is the sum of the cash flows discounted back to
the present, i.e. we wish to

$$\text{maximize } \sum_{j=1}^{N} \sum_{t=1}^{T} a_{jt}x_j/(1+r)^t.$$

The selections are constrained so that the sum of the expenditures does not
exceed the ceiling on expenditure set for each year:

$$\sum_{j=1}^{N} c_{jt}x_j \leq C_t \text{ for } t = 1, 2, ..., T.$$

Finally the problem variables are constrained as:

$$0 \leq x_j \leq 1, x_j \text{ integer.}$$

If it is possible to think in terms of partial acceptance of a project then x_j
may be regarded as the fraction of a project accepted and the same formulation applies without the integer constraints on the x_j quantities. However,
in practice the cash inflows a_{jt} may not be simply proportional to the fraction
of the project accepted, but depend on the size of the project through economies of scale. If the a_{jt} quantities depended proportionally on the magnitude
of the x_j quantities it would lead to a quadratic objective function, but the
interrelationship may be more complex than this. Yet another possibility
would be for the x_j to assume values greater than unity reflecting the desirabil-
ity of taking on multiples of attractive projects. But again this would not fit

within the proposed framework as returns to scale are not constant.

Weingartner makes a further point about this basic model which is worth noting. It is important to realize that the capital outlays c_{jt} are not the same as the cash flows a_{jt} for corresponding periods with sign reversed. Weingartner states that when budgets are used for control as well as planning purposes the c_{jt} quantities are to be controlled by the expenditure ceilings on capital account. The cash inflows resulting from project adoption should not be made available for reinvestment without first passing through the proper control channels and indeed may not be reinvested in the same project.

Another viewpoint on the programming of investment projects is expressed in Weingartner's 'basic horizon model'. This is more general than the previous model. Here the objective is to select investment projects and choose the cash investments and borrowings outside the firm so as to maximize the firm's value in some future year—'the horizon'. Some new factors and notation needs to be added to the basic model described above. The quantities a_{jt}, x_j are used as before and T is the horizon year. Let A_j be residual value of all physical assets acquired up to the horizon year T from project j. Also let v_t be the amount of cash available in year t which is invested outside the firm in year t at a rate of interest r_t and let w_t be the amount borrowed in year t at an interest cost r_t'. v_t and w_t are additional problem variables. Finally let D_t be the cash available from the current assets of the firm in year t before project selection takes place. This supply of cash could come for instance from projects already in existence. Then the value of the firm at the horizon is the residual assets of the projects plus the cash invested less the borrowings in year T. Therefore we must choose x_j, v_t and w_t to

$$\text{maximize} \sum_{j=1}^{N} A_j x_j + v_T - w_T.$$

The constraints express the requirement that the cash outflow should not exceed the cash inflow. In year 1 the cash outflow is v_1 and the cash inflow is the borrowing plus the returns from all projects

$$\sum_{j=1}^{N} a_{1j} x_j.$$

Thus for year 1, the cash flow constraint is

$$-\sum_{j=1}^{N} a_{1j} x_j + v_1 - w_1 \leq D_1$$

as D_1 is the cash available. For subsequent years we need to take account of borrowing or the interest earnings from the previous years giving the constraints

$$-\sum_{j=1}^{N} a_{jt} x_j - (1+r_{t-1})v_{t-1} + v_t + (1+r_{t-1}')w_{t-1} - w_t \leq D_t, \text{ for } t = 2, \ldots, T.$$

The extra problem variables v_t and w_t must be non-negative.

This model has been extended by Weingartner to incorporate dividend policies. Typical practical requirements may be that dividends should be non-decreasing in time, and that at the horizon, the value of remaining assets, physical and financial, should be sufficiently large to maintain in the future the dividend rate attained at the horizon. The following example outlines a practical example investigated by Chambers for the allocation of funds to investment projects where the allocation can be made over a period of years.

Example 8.2

A firm is evaluating the best way to allocate funds to 11 possible projects over a period of years. The projects are of three kinds: there are five miscellaneous small projects, three possible major investments and three projects such as repairs and maintenance for which the outlays are charged to expense rather than being capitalized. Table 8.2 describes these projects giving rates of return, possible profiles of the contribution over time to earnings before tax and the number of years for which they contribute to earnings.

Table 8.2

Project number	Rate of return Y	Profile of contribution over time to earnings before tax	No. of years which project contributes to earnings before tax	Nature of project outlays as proportion of total outlay
Miscellaneous investments				
1	11	Steady	5	Plant 100%
2	13	Rising	5	Plant 100%
3	14	Steady	10	Plant 50% Buildings 50%
4	10	Steady	5	Plant 100%
5	8	Falling sharply	5	Plant 100%
Major investments				
6		Steady	10	Plant 80% Buildings 20%
7		Rising to peak, then falling	10	Plant 60% Buildings 25% Stocks 15%
8		Steady	10	Plant 60% Buildings 25% Stocks 15%
Projects charged to expense				
9		Steady	4	
10		Steady	2	
11		Once for all	1	

The firm decides in advance in which years the projects may be started. It may well examine a number of alternative schemes. The problem variables then become the amounts of the disposable funds which should be allocated in each year to each project.

The objective is to maximize the present value of stockholders' dividends up to some horizon year T plus the terminal value at the end of the period. If D_t denotes the dividends declared at the end of period t and W_T is the part of the terminal value of the firm in year T contributed by projects started in the interval $(0, T)$ the objective function is

$$\sum_{t=0}^{T} D_t/(1+0{\cdot}07)^t + W_T/(1+0{\cdot}07)^T$$

where a discounting rate of 7 per cent has been assumed.

Chambers considered four types of practical constraints. First, the firm's policy was to increase published profits from year to year at a rate of 5 per cent. Secondly, in order to retain the confidence of the firm's creditors, it was desirable to keep the ratio of current assets to current liabilities, as reported on the balance sheet, at least equal to the value 3. Thirdly, it was required policy to maintain the dividend payment at one-third of earnings after tax. Finally, a performance measure, the ratio of earnings before tax or depreciation to total assets less current liabilities, was not to fall below $0{\cdot}15$.

The details of the way in which the objective function and the restrictions are expressed in terms of the outlays on the different projects are discussed in the reference by Chambers. The reference gives useful information on the way in which a practical case was organized.

8.6 Planning of advertising campaigns: static problems

Investigations into the use of optimization techniques for planning advertising campaigns are still at an early stage of development. The principal difficulty is to assess correctly the probabilistic response of a market to alternative advertising schemes. However, studies have been made which include various simplifying assumptions and these have led to models for advertising plans which have been used in practice. One useful feature in this field is the readily available data on the size of the populations which are exposed to advertisements such as the readership of the different newspapers or viewers of television programmes.

The studies fall broadly into two categories called static and dynamic problems. In the static problems the objective is to evoke a maximum response in terms of audience coverage or number of impacts, the time of attainment being unimportant. The dynamic campaigns are concerned with generating and maintaining a specific awareness level in a predetermined target population throughout a certain period; they are considered in the next section.

We will now consider the related problems of maximizing the coverage and the number of impacts in a static advertising campaign. The coverage is defined as the proportion of the population seeing at least one advertisement. The impact is defined as the number of times an advertisement is seen. The coverage and impact are therefore measures of dispersion and intensity respectively. They are examined in the following example.

Example 8.3

Suppose we have an amount of money C which we intend to use on an advertising campaign. We consider a number N of media such as newspapers and magazines in which the advertisement may be placed. It is assumed that each publication should carry the advertisement exactly once. The problem is to determine what size of advertisement should be placed in each of the media. (We are not concerned here with the contents of the advertisement.) It is assumed that the cost of inserting an advertisement is proportional to its size and that c_i is the cost of a full-page advertisement.

Let x_i denote the size of the advertisement measured as a proportion of a page which is to be inserted in medium i. These are our problem variables. We assume that the number of readers who see the advertisement in medium i is based on a square root law, i.e. the number is proportional to $\sqrt{x_i}$ and we denote the constant of proportionality by a_i. We also need to consider the proportion of readers who see the advertisement in more than one medium and assume that the number who see the advertisement in media i and j is $a_{ij}\sqrt{x_i}\sqrt{x_j}$. It will be assumed that a reader sees the same advertisement at most twice.

We can now work out the average number of impacts per person and the coverage. The probability that a particular person sees the advertisement in medium i is $a_i\sqrt{x_i}$. Therefore, as the advertisement appears once in each medium the mean number of impacts per person is

$$F(x_1, x_2, ..., x_N) = \sum_{i=1}^{N} a_i\sqrt{x_i}.$$

The coverage measuring the total number of people who see the advertisement (one or more times) must take account of the possibility of a person seeing the advertisement twice. If the advertisement was placed in two media, numbered 1 and 2 the proportion of the target population who see the advertisement only once is

$$a_1\sqrt{x_1} + a_2\sqrt{x_2} - 2a_{12}\sqrt{x_1}\sqrt{x_2}$$

as the proportion who see the advertisement in mediums 1 or 2 includes the proportion who see it twice. Thus the proportion who see it at least once is this quantity together with the proportion who see it exactly twice, i.e.

$$a_1\sqrt{x_1} + a_2\sqrt{x_2} - a_{12}\sqrt{x_1}\sqrt{x_2}.$$

Thus the coverage is given by

$$H(x_1, x_2, ..., x_N) = \sum_{i=1}^{N} a_i \sqrt{x_i} - \sum_{i=1}^{N} \sum_{\substack{j=1 \\ j>i \\ j \neq i}}^{N} a_{ij} \sqrt{x_i} \sqrt{x_j}.$$

The optimization problem is to maximize the impacts or the coverage, i.e. to maximize $F(x_1, x_2, ..., x_N)$ or $H(x_1, x_2, ..., x_N)$ subject to the following constraints. First the x_i must lie between zero and unity as they represent a proportion. Therefore

$0 \leqq x_i \leqq 1$, for $i = 1, ..., N$.

Secondly there is a maximum cost restriction on the funds available:

$$\sum_{i=1}^{N} c_i x_i \leqq C.$$

Although the objective function is an awkward expression these optimization problems can be solved by standard methods. It might be easier to transform the problem variables by writing

$x_i = y_i^2$

thus obtaining a linear objective function in the first case and a simple quadratic function in the second, subject to a single quadratic constraint, as the first constraint could still be written as

$0 \leqq y_i \leqq 1$.

This could readily be solved by a non-linear optimization technique.

8.7 Dynamic campaigns

The organization of a dynamic campaign is more difficult. Here we wish to generate and maintain a specific level of awareness in a predetermined target population throughout a certain period. For this purpose it is necessary to consider how people forget advertisements as well as how they see the advertisement in the first place. It is also necessary to introduce the notion of a measure of awareness. We will follow the definitions and development of this subject as proposed by Lee.

First it is necessary to consider how the readership of a single issue accumulates with time. For a publication which takes place on day 1 in medium i we assume that the increment in readership of the publication on day t can be expressed by a formula of the form

$r_i(t) = a_i(1 - p_i)p_i^{t-1}.$

This measure steadily decreases as t increases, which is what would be expected. The formula may be interpreted as giving the probability that a person who reads the publication will do so for the first time on day t. It also presumes that a 100 per cent readership a_i is only obtained after the lapse of a very long period. We will retain the rule assumed for static campaigns that the likelihood of someone seeing the advertisement is proportional to its size, and therefore the readership of the advertisement (as distinct from the publication) on day t is

$$a_i(1-p_i)p_i^{t-1}\sqrt{x_i}$$

if the size of the advertisement is x_i.

Next it is necessary to consider how people forget advertisements. The awareness level at any point in time is the percentage of the target population who remember the advertisement at that time. For an advertisement appearing on day 1 the percentage of the population remembering the advertisement on subsequent days steadily falls. Table 8.3 gives an example showing how the

Table 8.3

Day after publication	1	2	3	4	5	6	7
Awareness level: percentage of population aware	12	10	7	4	1	$\frac{1}{2}$	0

awareness levels may fall off over a seven day period. In general, we will assume the following rule of forgetting: the proportion of those people who saw an advertisement on day t and who remember it during a subsequent day d is

$$mq^{d-t}$$

where m and q are two memory parameters with $q<1$. Thus the proportion of the readership of medium i who are aware on day d of the advertisement appearing on day 1 is

$$e_i(d) = \sum_{t=1}^{d} a_i mq^{d-t}(1-p_i)p_i^{t-1}\sqrt{x_i}.$$

We can now present the basic optimization problem. We wish to determine when to insert advertisements in the media so that a predicted level-of-awareness pattern is generated at a minimum cost. For example we might require the level of awareness over a 6-week period to be as illustrated in Table 8.4. Assuming that the sizes of the advertisements are known, the

Table 8.4

Week Number	1	2	3	4	5	6
Minimum level of awareness (percentage of population aware)	35	35	25	20	15	15

problem variables are the days on which they are to appear in the various media. Let $y_i(j) = 1$ if the issue of medium i on day j carries an advertisement, and let $y_i(j) = 0$ if it does not. The problem variables can thus take on the values zero and one. We assume there are N media and the period is set for n days.

If the cost of inserting an advertisement is proportional to its size, and c_i is the constant of proportionality for medium i, the total cost of the advertising schedule is

$$\sum_{i=1}^{N} \sum_{j=1}^{n} y_i(j)c_i x_i.$$

The constraints are to ensure that the minimum level of awareness is attained. Let L_k be the required level on day k. Then for day 1 we require

$$\sum_{i=1}^{N} y_i(j)e_i(1) \geqq L_1$$

for day 2

$$\sum_{i=1}^{N} \sum_{j=1}^{2} y_i(j)e_i(3-j) \geqq L_2$$

and in general for day k

$$\sum_{i=1}^{N} \sum_{j=1}^{k} y_i(j)e_i(k+1-j) \geqq L_k.$$

Finally we require

$$y_i(j) = 1 \text{ or } 0.$$

This is an integer-linear programming problem, and may be solved by linear programming combined with the branch and bound method. Sometimes special techniques can be used. In particular, Kolesar has studied the relationship between some media scheduling problems and the knapsack problem which lead on to a direct solution by branch and bound methods.

REFERENCES

Chambers, D. 1967. Programming the allocation of funds subject to restrictions on reported results. *Opl. Res. Quat.*, **18**, 407.
Ellis, D. M. 1966. Building up a sequence of optimum media schedules. *Opl. Res. Quat.*, **17**, 413.
Kolesar, P. J. 1968. A remark on the computation of optimum media schedules. *Opl. Res. Quat.*, **19**, 73.
Lee, A. M. 1963. Decision rules for media scheduling: dynamic campaigns. *Opl. Res. Quat.*, **14**, 365.
Lee, A. M. and Burkhart, A. J. 1960. Some optimization problems in advertising media planning. *Opl. Res. Quat.*, **11**, 113.
Mao, J. C. T. and Sarndal, C. E. 1966. A decision theory approach to portfolio selection. *Man. Sci.*, **12** (8), B323.

Naslund, B. and Whinston, A. 1962. A Model of multi-period investment under uncertainty. *Man. Sci.*, **9**, 184.
Robichek, A. A. Teichroew, D. and Jones, J. M. 1965. Optimal short term financing decision *Man. Sci.*, **12**, 1.
Sharpe, W. F. 1967. A linear programming algorithm for mutual fund portfolio selection. *Man. Sci.*, **13** (7), 499.
Taylor, C. J. 1963. Some developments in the theory and application of media scheduling methods. *Opl. Res. Quat.*, **14**, 291.
Weingartner, H. M. 1967. Optimal timing of bond refunding. *Man. Sci.*, **13** (7), 511.
Weingartner, H. M. 1966. Criteria for Programming Investment Project Selection. *J. Indust. Econ.*, **15**.

Exercises on Chapter 8

1 A financial secretary of a firm wants to invest £100,000 so as to maximize the yield. He considers the following types of securities:

Type	1	2	3	4	5	6
Yield	3·5%	2·5%	3%	4·5%	5%	4%

He is not completely free in his choice, because it is the firm's policy that at least 35 per cent of the whole amount should be invested in units 1 or 2 and that not more than 40 per cent should be invested in either types 3 or 4 combined or types 5 and 6 combined. How should the money be invested.

2 An investor has £150 and is considering investing in three securities. The price of the securities are £10, £5, and £5 respectively and their mean rate of return is 6, 4 and 8 respectively. The returns on the securities are uncorrelated and their variances are 1.5, 1 and 2.5 respectively. Formulate the problem of maximizing the return assuming that the investor wishes to ensure that the variance of his investment is less than 1000. Show that it is not the best policy to buy as much as possible of the security with the largest rate of return.

3 An advertiser considers how best to spend £600 on placing advertisements in four media. The response value measured in some convenient unit varies between media and it also depends on how expensive is the style of advertising in the particular medium. The response for the various expenditures are shown in the Table 8.5.

Table 8.5

		Medium			
		1	2	3	4
	0	0	0	0	0
	1	4	1	2	3
	2	4	2	2	4
Expense in 3	3	6	3	4	6
£100	4	6	4	4	6
	5	7	7	6	7
	6	7	9	8	8

Determine how the money should be allocated to obtain the greatest response value.

4 Use the formulae described in Section 8.7 to calculate the levels of awareness resulting from a single quarter-page advertisement in a medium with the following characteristics: total readership 1000, of which nine-tenths read the advertisement on day 1 and the estimated memory parameters are $m = 1$, $q = 0.6$.

Answers to exercises

Answers 1

1 For any value of a, the sum $S(a)$ of the squared differences between the observed and predicted values of a is

$$S(a) = 0 + (2-a)^2 + (2-4a)^2 + (5-9a)^2.$$

The equation $\dfrac{dS}{da} = 0$ gives the value of a as $\frac{55}{98}$, and provides the predicted values for $x = 0, 1, 2, 3$ as shown in Tables A12.1.

Table A1.1

x	0	1	2	3
Observed	0	2	2	5
Predicted	0	$\frac{55}{98}$	$2\frac{12}{49}$	$5\frac{5}{98}$

The greatest deviation occurs at $x = 1$, where the deviation is $1\frac{43}{98}$.

2 The sum $S(a, b, c)$ of the squared deviations between the observed and predicted values is the function

$$S(a, b, c) = \sum_{i=1}^{N} (y^{(i)} - ae^{-bx_1^{(i)}} - cx_2^{(i)})^2.$$

If we differentiate this function with respect to a and equate the resulting function to zero we obtain the equation:

$$-2 \sum_{i=1}^{N} e^{-bx_1^{(i)}}(y^{(i)} - ae^{-bx_1^{(i)}} - cx_2^{(i)}) = 0.$$

Differentiating with respect to b we obtain the equation:

$$2 \sum_{i=1}^{N} a x_1^{(i)} e^{-bx_1^{(i)}} (y^{(i)} - a e^{bx_1^{(i)}} - c x_2^{(i)}) = 0.$$

Finally differentiating with respect to c we obtain the equation

$$-2 \sum_{i=1}^{N} x_2^{(i)} (y^{(i)} - a e^{-bx_1^{(i)}} - c x_2^{(i)}) = 0.$$

However it would be very difficult to solve these three equations to obtain a, b and c. It would be better to work with the non-linear function $S(a, b, c)$ and minimize this directly by a descent or direct search method.

3 If the proposed function relationship is

$$y = f(x)$$

and N observations are taken $(y^{(i)}, x^{(i)})$ for $i = 1, ..., N$, the least squares procedure minimizes

$$\sum_{i=1}^{N} (y^{(i)} - f(x^{(i)}))^2.$$

If the deviations were not squared but simply summed

$$S = \sum_{i=1}^{N} (y^{(i)} - f(x^{(i)}))$$

we could get a very small value of S, despite a poor correspondence between the $y^{(i)}$ and $f(x^{(i)})$, provided the positive and negative deviations cancelled out. This difficulty could be overcome by taking the modulus of the deviations to give the sum as,

$$S = \sum_{i=1}^{N} | y^{(i)} - f(x^{(i)})|.$$

But it might be awkward to minimize this function as the derivatives are not continuous. Besides it may be an advantage to stress that the larger deviations should be minimized. The squaring of the deviations tends to achieve this, but it would be further stressed if we minimized the sum of the fourth powers as

$$S = \sum_{i=1}^{N} (y^{(i)} - f(x^{(i)}))^4.$$

If this criterion was very important it might be worth minimizing the maximum deviation as

$$\text{minimize } \{ \max_{1 \leq i \leq N} (y^{(i)} - f(x^{(i)})) \}.$$

Answers 2

1 The two constraints on the controls restrict the feasible solution to lie in the rectangle *ABCD* of Fig. A2.1. Initially the additional constraint is the

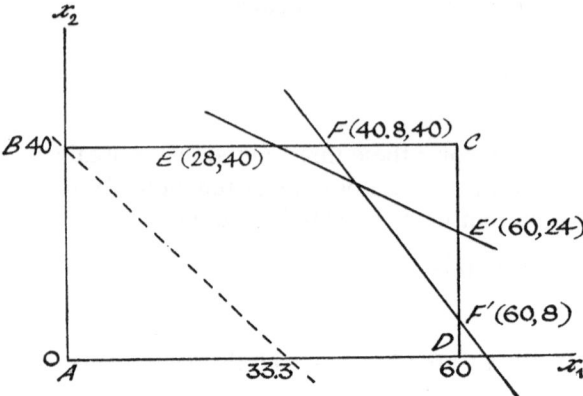

Fig. A2.1

line *EE'* but after the disturbance it becomes *FF'*. The slope of the objective function is indicated by the dotted line. It is clear graphically by considering the slope of the dotted line (representing a particular value of the objective function) that before the disturbance the optimum control settings are at *E'*, i.e. $x_1 = 60$, $x_2 = 24$ with profit 430 units. After the disturbance the the optimum settings correspond to *F*, i.e. $x_1 = 40.8$, $x_2 = 40$ with profit 394·8.

2 The process is illustrated in the diagram

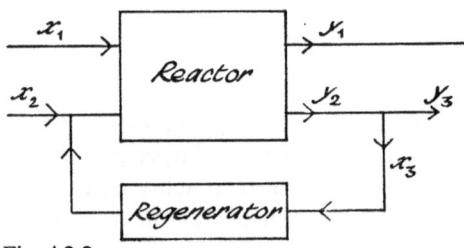

Fig. A2.2

x_1, x_2 are the flow rates of the feed lines from raw material stocks.
x_3 is the amount of by-product which is re-cycled.
y_1, y_2 are the output rates of the product and by-product.
y_3 is the amount of by-product which is sold.
p_1, p_2 are the selling prices of the product and by-product per unit and c_1, c_2 are the costs of the two raw materials. c_3 is the cost of regeneration per unit

time. These are the disturbance variables. The total profit from the process which is to be maximized is

$$p_1 y_1 + p_2 y_3 - c_1 x_1 - c_2 x_2 - c_3 x_3.$$

The process constraints require that

$$y_1 = h(x_1, x_2 + x_3).$$

A total mass balance requires that

$$x_1 + x_2 + x_3 = y_1 + y_2.$$

Also the total amount of by-product must equal the amount sold and regenerated, i.e.

$$y_2 = y_3 + x_3.$$

These three equations enable the quantities y_1, y_2, y_3 to be expressed in terms of the control variables x_1, x_2, x_3. Finally the inlet flow rates must lie between limits say l_i and l_i' so that

$$l_1 \leqq x_1 \leqq l_1' \quad l_2 \leqq x_1 + x_3 \leqq l_2'.$$

Also the recycled product must satisfy the constraints

$$0 \leqq x_3 \leqq y_2.$$

This completes the optimization problem.

3 The problem variables are the quantities of A, C, S and I which go into N, H and P. Denote by A_N, C_N, S_N and I_N, the quantities of A, C, S and I going into product N and similarly define A_H, C_H, S_H and I_H and A_P, C_P, S_P and I_P. By performing a mass balance the quantity of product N is then $A_N + C_N + S_N + I_N$ and similarly for H and P. Therefore the objective function to be maximized is:

$$5(A_N + C_N + S_N + I_N) + 6(A_H + C_H + S_H + I_H) + 4(A_P + C_P + S_P + I_P).$$

As all the product is used we require that the total of each component used per unit time is equal to the supply rate:

$$A_N + A_H + A_P = 38$$
$$C_N + C_H + C_P = 25$$
$$S_N + S_H + S_P = 40$$
$$I_N + I_H + I_P = 14.$$

To achieve the performance numbers required for the aviation gasoline, we require for N that

$$107 \frac{A_N}{(A_N + C_N + S_N + I_N)} + 93 \frac{C_N}{(A_N + C_N + S_N + I_N)} + 87 \frac{S_N}{(A_N + C_N + S_N + I_N)}$$

$$+ 108 \frac{I_N}{(A_N + C_N + S_N + I_N)} \geqq 91$$

H

which may be written as

$$107A_N + 93C_N + 87S_N + 108I_N \geqq 91(A_N + C_N + S_N + I_N).$$

Similarly,

$$107A_H + 93C_H + 87S_H + 108I_H \geqq 100(A_H + C_H + S_H + I_H).$$

Finally to ensure that the vapour pressures do not exceed 7 lb in^2;

$$5A_N + 8C_N + 4S_N + 20I_N \leqq 7(A_N + C_N + S_N + I_N)$$
$$5A_H + 8C_H + 4S_H + 20I_H \leqq 7(A_H + C_H + S_H + I_H).$$

It is clear that this is a linear programming problem and can be solved by the simplex method. The actual solution to this problem is given in the table below.

Table A2.1

	N	H	P
A	0·120	37·880	0
C	0	25·000	0
S	29·116	10·884	0
I	6·738	6·416	0·846

4 The transfer matrix for converting products 1 and 2 into 2, 3 and 4 is given in Table A2.2.

Table A2.2

Products

	b	1	2	3	4
Transfer 1		-1	$\frac{1}{4}$	$\frac{1}{2}$	$\frac{1}{4}$
2			-1	$\frac{2}{5}$	$\frac{3}{5}$

The problem variables are

x_i = the amount in tons of product i to be put through the grinder ($i = 1, 2$)

s_i = quantity in tons of product i made for sales

d_i = quantity in tons of product i made to be dumped.

Total profit is

$$F(X, S, D) = 7s_1 + 10s_2 + 16s_3 + 25s_4$$
$$- 5(d_1 + d_2 + d_3 + d_4)$$
$$- 5x_1 - 10x_2.$$

The total amount produced must equal the total amount sold or dumped.

$$10 \ -x_1 \qquad\quad = s_1 + d_1$$
$$15 \ +\tfrac{1}{4}x_1 - x_2 \ = s_2 + d_2$$
$$17 \!\cdot\! 5 + \tfrac{1}{2}x_1 + \tfrac{2}{5}x_2 = s_3 + d_3$$
$$7 \!\cdot\! 5 + \tfrac{1}{4}x_1 + \tfrac{3}{5}x_2 = s_4 + d_4.$$

All variables must be positive and

$$s_1 \leqq 10, \quad s_2 \leqq 24, \quad s_3 \leqq 15, \quad s_4 \leqq 18.$$

This is a linear programming problem.

5 The differential equations of the reaction may be expressed as

$$\frac{dx}{dt} = -K_1 x$$

$$\frac{dy}{dt} = K_1 x - K_2 y$$

$$x + y + z = 1.$$

Also the kinetic reaction coefficients are expressed as functions of pressure

$$K_1 = f_1(P) = C_1 \cdot P$$
$$K_2 = f_2(P) = C_2/P$$

where P is the pressure at time t.

To use dynamic programming a skew grid would be determined in the (x, y) plane with slopes S_1 and S_2 (neither set of grid lines parallel to an axis), and we would rewrite the differential equations as difference equations relating node points (x_j, y_j) and (x_{j+1}, y_{j+1}) as

$$\frac{x_{j+1} - x_j}{\Delta t} = -K_1 x_j$$

$$\frac{y_{j+1} - y_j}{\Delta t} = K_1 x_j - K_2 y_j.$$

To determine the time to move from one node to the next in the directions in a grid imposed over the (x, y) space, we calculate

$$\Delta t = - \frac{x_{j+1} - x_j}{K_1 x_j}$$

The values of K_1 and K_2 are determined by the equations

$$K_1 = C_1 \left(\frac{(P_{j+1} + P_j)}{2} \right)$$

$$K_2 = C_2 \left(\frac{2}{P_{j+1} + P_j} \right)$$

where C_1 and C_2 are constants of proportionality.

The pressure P_{j+1} at node $(j+1)$ is determined by the equation

$$\frac{y_{j+1}-y_j}{x_{j+1}-x_j} = \frac{K_1 x_j - K_2 y_j}{-K_1 x_j}$$

$$= \frac{\dfrac{C_1}{2}(P_{j+1}+P_j)x_j - C_2\left(\dfrac{2}{P_{j+1}+P_j}\right)y_j}{-\dfrac{C_1}{2}(P_{j+1}+P_j)x_j}$$

as all quantities are known except P_{j+1}. This enables us to determine K_1 and hence calculate Δt. The time taken to move from node j to node $(j+1)$ will be $\Delta T_1(P_{j+1})$ or $\Delta T_2(P_{j+1})$ according to whether we are proceeding in the direction S_1 or S_2. We can now use the same dynamic programming recurrence relations as were used in Section 13.7 to determine the optimum:

$$f(x_j, y_j, P_j) = \min_{L_1 \leq P_J \leq L_2} \begin{cases} (\Delta T_1(P_j)+f_{k-1}(x_i, y_i, P_i) \\ \Delta T_2(P_j)+f_{k-1}(x_l, y_l, P_l) \end{cases}$$

with the additional limitations on the pressure variation included.

6 Dynamic programming can be used to solve this problem. Let $f_n(p_n)$ be the minimum energy required to compress a gas to pressure p_n from pressure p_0 in n stages.

Then the recurrence relations are

$$f_n(p_n) = \min_{p_0 \leq p_{n-1} \leq p_n} \left[\left(\frac{p_n}{p_{n-1}}\right)^q + f_{n-1}(p_{n-1})\right]$$

as the pressure p_{n-1} must lie between p_0 and p_n, and the energy for given p_n is the sum of the energies required at stage n and the minimum energy required over the first n stages.

In a three-stage system we first determine $f_2(p_2)$ where p_2 is not yet known. This is found by determining the pressure p_1 in the range (p_0, p_2) to minimize

$$h(p_1) = \left(\frac{p_2}{p_1}\right)^q + \left(\frac{p_1}{p_0}\right)^q$$

$$\frac{dh}{dp_1} = \frac{q}{p_1}\left\{\left(\frac{p_1}{p_0}\right)^q - \left(\frac{p_2}{p_1}\right)^q\right\}.$$

This derivative is negative for $p_1 = p_0$ and positive for $p_1 = p_2$ so that the minimum must lie between these limits. Equating to zero gives

$$p_1 = (p_0 \cdot p_2)^{\frac{1}{2}}$$

and

$$f_2(p_2) = 2\left(\frac{p_2}{p_0}\right)^{q/2}.$$

In the next recurrence relation we need to minimize

$$h(p_2) = \left(\frac{p_3}{p_2}\right)^q + 2\left(\frac{p_2}{p_0}\right)^{q/2}$$

with respect to p_2, $p_0 \leqq p_2 \leqq p_3$. Differentiating gives

$$\frac{dh}{dp_2} = \frac{q}{p_2}\left\{\left(\frac{p_2}{p_0}\right)^{q/2} - \left(\frac{p_3}{p_2}\right)^q\right\}$$

and equating to zero gives

$$p_2 = (p_0 \cdot p_3^2)^{\frac{1}{3}}.$$

Hence the intermediate pressures between p_0 and p_3 (which are given) are

$$p_1 = (p_0^2 p_3)^{\frac{1}{3}}$$
$$p_2 = (p_0 p_3^2)^{\frac{1}{3}}.$$

7 Dynamic programming can again be used to solve this problem. Define $f_n(C_n)$ as the minimum total holding time required over n tanks to obtain concentration C_n. Then the holding time at the nth stage can be expressed as:

$$h_n = \frac{C_n - C_{n-1}}{K_1 - (K_1 + K_2)C_n}.$$

We can now write down the recurrence relationship as

$$f_n(C_n) = \min_{C_{n-2} \leqq C_{n-1} \leqq C_n} [h_n + f_{n-1}(C_{n-1})]$$

$$= \min_{C_{n-2} \leqq C_{n-1} \leqq C_n}\left[\frac{C_n - C_{n-1}}{K_1 - (K_1 + K_2)C_n} + f_{n-1}(C_{n-1})\right]$$

and solve it for $n = 2, 3, 4, ..., N$, where

$$f_1(C_1) = \frac{C_1 - C_0}{K_1 - (K_1 + K_2)C_2}.$$

For a two-tank system we have

$$f_1(C_1) = \frac{C_1 - C_0}{K_1 - (K_1 + K_2)C_2}$$

and

$$f_2(C_2) = \min_{C_0 \leqq C_1 \leqq C_2}\left[\frac{C_2 - C_1}{K_1 - (K_1 + K_2)C_2} + \frac{C_1 - C_0}{K_1 - (K_1 + K_2)C_1}\right].$$

Differentiating, with respect to C_1 we require

$$\frac{1}{K_1 - (K_1 + K_2)C_1} + \frac{(K_1 + K_2)(C_1 - C_0)}{(K_1 - (K_1 + K_2)C_1)^2} - \frac{1}{K_1 - (K_1 + K_2)C_2} = 0.$$

(It should be noted that the derivative is negative when $C_1 = C_0$ and positive when $C_1 = C_2$ so that the minimum must lie between C_0 and C_2.) To solve this equation for C_1 set

$$\frac{K_1}{K_1+K_2} = C_e, \quad C_e - C_i = x_i \quad \text{for} \quad i = 0, 1, 2.$$

Then

$$\frac{1}{(K_1+K_2)x_1} + \frac{x_0-x_1}{(K_1+K_2)x_1^2} - \frac{1}{(K_1+K_2)x_2} = 0$$

giving $x_0 x_2 = x_1^2$

i.e. $(C_1 - C_e) = \{(C_e-C_0)(C_e-C_2)\}^{\frac{1}{2}}.$

Now

$$h_1 = \frac{1}{(K_1+K_2)}\left(\frac{C_1-C_0}{C_e-C_1}\right) = \frac{1}{(K_1+K_2)}\frac{(C_e-C_0)-(C_e-C_1)}{C_e-C_1}$$

$$= \frac{1}{(K_1+K_2)}\left[\left(\frac{C_e-C_0}{C_e-C_2}\right)^{\frac{1}{2}} - 1\right].$$

Similarly

$$h_2 = \frac{1}{(K_1+K_2)}\left(\frac{C_2-C_1}{C_e-C_2}\right) = \frac{1}{(K_1+K_2)}\left[\left(\frac{C_e-C_0}{C_e+C_2}\right)^{\frac{1}{2}} - 1\right].$$

Thus the holding times h_1 and h_2 are equal.

8 The backwards calculation of dynamic programming can be used to solve this problem. Define $f_i(y)$ as the maximum profit obtainable from the last $(i+1)$ products if a total amount of material y enters the ith last screen. The profit obtained at this stage is

$$(1-z_i)p_i y$$

and the cost of the separation is

$$k_i y \log (1-z_i).$$

Therefore the recurrence relations are

$$f_i(y) = \max_{0 \leq z_i y \leq l_i} \{(1-z_i)p_i y - k_i y \log (1-z_i) + f_{i-1}(z_i y)\}$$

The maximization is obtained over z_i for various values of y and for $i = 2, 3, 4$ where $f_1(y) = p_1 y$. The maximum profit is $f_3(l_1+l_2+l_3+l_4)$ and the z_i values for which the maximum is obtained at each stage can be determined. By noting how much of the totals l_i of each product in the feed are in fact passed on to the next screen we can determine the composition of the products in terms of the grades of materials.

9 The total costs for the supply distribution $(x_1, x_2, ..., x_N)$ are

$$F(x_1, x_2, ..., x_N) = \sum_{i=1}^{N} f_i(x_i).$$

The supply must equal the demand so that

$$\sum_{i=1}^{N} x_i - L(x_1, x_2, ..., x_N) = D.$$

This equality constraint can be accommodated in a revised objective function with a Lagrange multiplier y as

$$H(x_1, x_2, ..., x_N, y) = \sum_{i=1}^{N} f_i(x_i) - y\left\{ \sum_{i=1}^{N} x_i - L(x_1, x_2, ..., x_N) - D \right\}.$$

The conditions for optimum operation are that the derivatives of H are zero, i.e.

$$\frac{\partial H}{\partial x_i} = \frac{\partial f_i}{\partial x_i} - y\left(1 - \frac{\partial L}{\partial x_i}\right) = 0$$

and

$$\sum_{i=1}^{N} x_i - L(x_1, x_2, ..., x_N) = D.$$

The first set of equations may be written as

$$\frac{\partial f_i}{\partial x_i}\left(1 - \frac{\partial L}{\partial x_i}\right)^{-1} = y \text{ for all } i, 1 \leq i \leq N$$

and these equations determine the optimum power supply distribution from the N generating sources.

Answers 3

1 Using Table 3.2, when jobs 1 and 4 have been loaded the current state of the machines is as illustrated. The completion times of these two jobs is

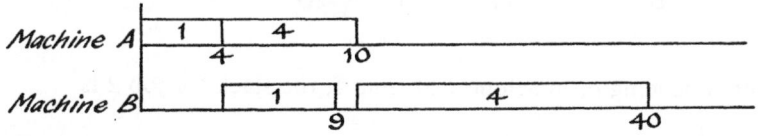

Fig. A3.1

therefore 9 and 40. Jobs 2, 3 and 5 have ordered processing times on machine A as 2, 4 and 30 so that the bound $T(E_r)$ is

$10+2+3$

$+10+2+4+1$

$+10+2+4+30+4$

$= 82.$

The times of the three excluded jobs on machine B are 1, 3 and 4 so that the bound $S(E_r)$ is worked out as

$40+1$

$40+1+3$

$40+1+3+4$

$= 133.$

There the bound on the total completion time is $9+40+133 = 182.$

2 The schedule obtained by using the shortest operation rule is shown in the Fig. A3.2 in which the numbers shown are the job numbers. The completion

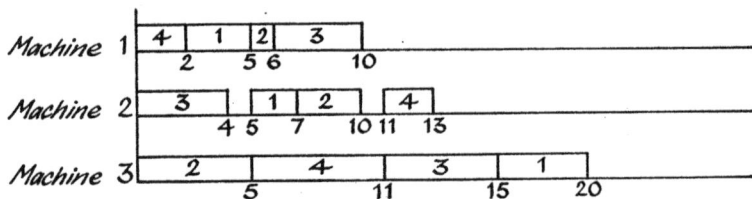

Fig. A3.2

times of the jobs are 20, 10, 15 and 13 respectively. Therefore the maximum job lateness is 6 and the total idle time is 2 units on machine 2. Using the least slack rule the schedule is as shown in Fig. A3.3. For example, when

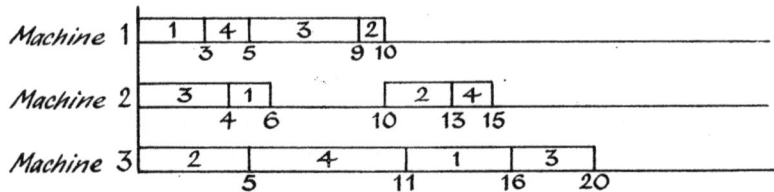

Fig. 3.3

jobs 2 and 3 are queueing on machine 1 at time 5, the slack for job 2 is

$20-5-1-3 = 11$

and the slack for job 3 is

$18-5-4-4 = 5$

so that job 3 is chosen.

The completion times of the jobs are 16, 13, 20, 15 and the maximum job lateness is 2, which is a considerable improvement on the shortest operation discipline. However, the machine idle time is 4 units which is worse.

3 The transformation determines the six start times as: 0, 0, 2, 5, 3, 0 and the resource usage is illustrated in the Fig. A3.4.

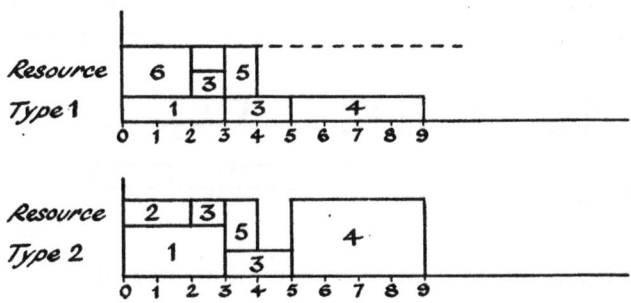

Fig. A3.4

4 The full set of data on the durations is given in Table A3.1. If the shortest

Table A3.1

	A	B	A	B
Job 1	6	6	3	3
Job 2	4	2	2	1
Job 3	2	4	1	2

operation rule is applied to this data the operation start times are determined as shown in Fig. A3.5 in which the numbers in the rectangles on the bar

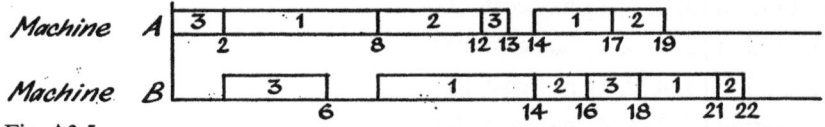

Fig. A3.5

chart are the jobs. The problem can be expressed quite simply in terms of a permutation. Let the operations of job j on machines A and B be denoted by the elements $(j, 1)$ to $(j, 4)$. The operations can be ordered as a list of 12 elements. The ordering can be constrained so that operation $(j, k+1)$ is placed after operation (j, k). Then the start times of the operations can be determined by taking each operation in turn and starting it as early as possible without disturbing the positions of the previously allocated operations. For example the schedule of operations achieved by the shortest operation rule in Fig. A3.5 can be obtained by applying this scheme to the permutation

$(3, 1), (2, 1), (3, 2), (3, 3), (2, 2), (1, 1), (3, 4), (2, 3), (1, 2), (1, 3), (2, 4), (1, 4).$

If we wish to minimize the maximum lateness with all due dates set to 19, job 1 can probably be improved in the schedule of Fig. A3.5 where it has a lateness of 6 units. We can try moving operation $(1, 1)$ to the second position in the above permutation, operation $(1, 2)$ to the fifth position, operation $(1, 3)$

to the eighth position and operation (1, 4) to the eleventh position to give
the permutation:

(3, 1), (1, 1), (2, 1), (3, 2), (1, 2), (3, 3), (2, 2), (1, 3), (3, 4), (2, 3), (1, 4), (2, 4).

This gives the schedule shown in Fig. A3.6 for which the maximum lateness

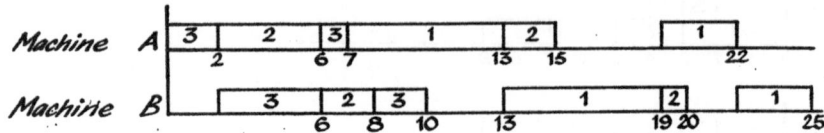

Fig. A3.6

is 3 units (for job 2) and is an improvement over the previous schedule.

5 Let p_j be the jth order to be processed. The permutation is:

$$[P] = [p_1, p_2, ..., p_N].$$

Let $x(i)$ denote the completion time of order i. Then the $x(i)$ are determined
recursively as:

$$x(p_j) = x(p_{j-1}) + S(p_{j-1}, p_j) + r(p_j) \cdot W(p_j) \text{ for } j = 1, N$$
$$x(p_0) = 0.$$

The total cost of machine time and completion times, is

$$F[P] = b \cdot x(p_N) + \sum_{j=1}^{N} C(p_j) \cdot \max (x(p_j) - T(p_j), 0).$$

The optimization problem is to determine a permutation $[P]$ to minimize $F[P]$.

The permutation [3, 2, 4, 1] has cost   72

The permutation [2, 3, 4, 1] has cost 100

The permutation [3, 4, 2, 1] has cost 119

The permutation [3, 2, 1, 4] has cost   80

Therefore the permutation [3, 2, 4, 1] is optimal with respect to adjacent
exchanges.

6 Table A3.2 shows the possible combinations.

Table A3.2

Combination

		1	2	3	4	5	6	7	8	9	10	11	12
	58	1	1										
	26			3	2	2	1	1	1				
Widths	24	1			1		2	1		3	2	1	
	23		1			1		1	2		1	2	3
Wastage		0	1	4	6	7	8	9	10	10	11	12	13

Let x_i be the number of reels of combination i. Then the constraints are

$x_1 + x_2 \geq 60$

$3x_3 + 2x_4 + 2x_5 + x_6 + x_7 + x_8 \geq 85$

$x_1 + x_4 + 2x_6 + x_7 + 3x_9 + 2x_{10} + x_{11} \geq 85$

$x_2 + x_5 + x_7 + 2x_8 + x_{10} + 2x_{11} + 3x_{12} \geq 50.$

Using slack variables x_{13}, x_{14}, x_{15}, x_{16}, the objective function is to minimize

$x_2 + 4x_3 + 6x_4 + 7x_5 + 8x_6 + 9x_7 + 10x_8 + 10x_9 + 11x_{10} + 12x_{11}$
$$+ 13x_{12} + 58x_{13} + 26x_{14} + 24x_{15} + 23x_{16}.$$

7 The application of the cutting rule will require 12 guillotine cuts, shown in Fig. A3.7 as numbers in circles.

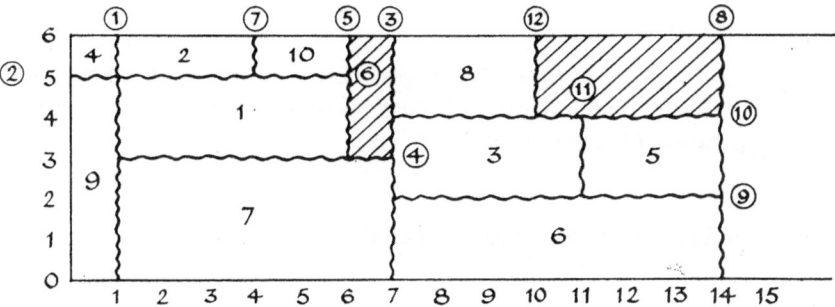

Fig. A3.7

The first and second cuts produce rectangle 9 and rectangle 4

The third and fourth cuts produce rectangle 7.

The fifth and sixth cuts produce rectangle 1.

The seventh cut produces rectangles 2 and 10 from the scrap.

The eighth and ninth cuts produce rectangle 6.

The tenth and eleventh cuts produce rectangles 3 and 5.

The twelfth cut produces rectangle 8.

8 Let N be the number of operations in the network, let M be the number of nodes and let $i(n)$, $j(n)$ for $n = 1$ to N be the start and end nodes of the kth operation where $i(n)$, $j(n)$ are numbers between 1 and M. Also let $d(i, j)$ be the duration of the operation from node i to node j. First we determine the earliest time at which each operation (i, j) can start. This is equal to the longest series of operations which must be performed before node n can be reached. Define $f_k(m)$ as the longest time it can take to reach node m in the network using series of at most k connected operations. Then starting with $f_0(m) = 0$ for all m, we can evaluate the recurrence relations

$$f_k(m) = \max_j \left[f_{k-1}(j) + d(j, m) \right] \text{ for } 1 \leq m \leq M \text{ and } k = 1, 2, \ldots$$

where the maximization extends over values of j for which there is an operation (j, m) in the network. When the recurrence relations have been evaluated up to an iteration $k = L$ for which no $f_k(i)$ value has altered we obtain the earliest start time of an operation out of node m as $f_L(m)$ and the minimum project time is $f_L(M)$ where M is the final node in the network.

We now find the latest time at which an operation (j, m) must be finished by finding the longest series of operations from node M to node m by doing a reverse calculation on the network. If this time is $f'_{L'}(m)$, then for any operation (i, j) we know that it cannot start until time $f_L(i)$ as the whole project starts at time 0 and we know that it must have started by time $f_L(M) - f'_{L'}(j) - d(i, j)$ for the whole project completion time to be unaffected. Hence the slack is measured as the difference between the latest and earliest start times:

$$f_L(M) - f'_{L'}(j) - d(i, j) - f_L(i).$$

In the example network of 6 nodes the longest path to node 3 is 6 units. The total project time is 18. The longest path back to node 5 is 5 units. Therefore the slack on operation $(3, 5)$ with duration 3 is

$$18 - 5 - 3 - 6 = 4.$$

9 Assume the indices of the jobs are numbered so that

$$T(j) \leq T(j+1).$$

Then the heuristic rule suggests ordering the jobs as 1, 2, ..., N. The completion time $C(j)$ of job j is

$$C(j) = \sum_{k=1}^{j} d(k)$$

and the maximum lateness to be minimized is

$$\max_{1 \leq j \leq N} \{\max (C(j) - T(j), 0)\}.$$

Suppose that we have a sequence in which two neighbouring jobs are not in due date order. Let us assume that these jobs are numbered 1 and 2, for which $T(1) < T(2)$. In the order $(2, 1)$ the completion times are

$$C(1) = D + d(1) + d(2),$$

$$C(2) = D + d(2),$$

where D is the finish time of the job prior to this pair of jobs and the maximum lateness is say X where

$$X = \max \{\max (D + d(1) + d(2) - T(1), 0), \max (D + d(2) - T(2), 0)\}$$

$$= \max (D + d(1) + d(2) - T(1), 0).$$

In the order (1, 2) the maximum lateness is Y where

$$Y = \max \{\max (D+d(1)-T(1m), 0), \max (D+d(1)+d(2)-T(2), 0)\}.$$

We need to show that

$$X \geq Y.$$

Firstly, if $X = 0$, $Y = 0$ because the first component in X is larger than either component in Y.

Secondly, if X is positive we require that

$$D+d(1)+d(2)-T(1) \geq \max (D+d(1)-T(1), 0)$$

i.e. $d(2) \geq 0$, which is true,

and $D+d(1)+d(2)-T(1) \geq \max (D+d(1)+d(2)-T(2), 0)$

i.e. $T(2)-T(1) \geq 0$, which is also true.

Therefore it is never a disadvantage to interchange a pair of jobs which are not in due time order and it may be an advantage. By repeated application of this result we obtain the result that the heuristic rule will reach an optimal sequence.

To demonstrate that this scheme will not necessarily lead to the best result if we are minimizing the maximum deviation from the due date rather than the maximum lateness, consider the two jobs numbered 1, 2 with durations 3 and 5 and due times 9 and 10, and the machine becoming empty at time 0. If the jobs are placed in due time order as 1, 2 they are completed at times 3 and 8 respectively and the maximum deviation between the due time and the completion time is max $(9-3, 10-8) = 6$. If the jobs are placed in the order (2, 1) their completion times are 5, 8 and the maximum deviation from the due time is max $(9-8, 10-5) = 5$ so that the maximum deviation is larger with the due time ordering.

10 Let us take three jobs 1, 2, 3 with durations on the machines A and B

$a(i)$ and $b(i)$ for $i = 1, 2, 3$.

If the jobs are arranged on machine A in the order 1, 3, 2 and on machine B in the order 2, 3, 1 the schedules will appear as illustrated in Fig. A3.8, if

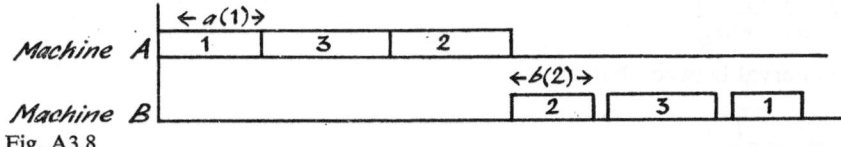

Fig. A3.8

each operation is taken in turn and loaded at its earliest possible start time. Then without any loss of total production time we can make the ordering on machine A the same as on machine B by making successive interchanges to the operations on machine A. Hence we can always regard the sequences on the two machines as the same.

Answers 4

1 The stock level will have the form shown in Fig. A4.1. If a production

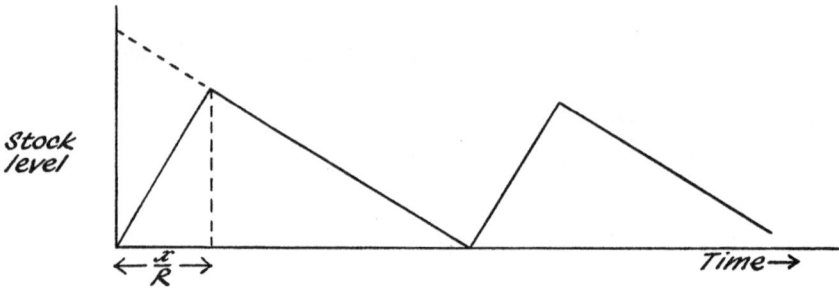

Fig. A4.1

run starts at time 0 stocks will build up at a rate $(R-r)$ until the batch is complete, and thereafter will decline at rate r per unit time. Let c_1 denote the cost of a set-up, and c_2 be the cost of storage per unit per unit time. If x denotes the batch size (and therefore $\dfrac{x}{r}$ is the interval between production runs) the storage will go on increasing for a time x/R to reach a maximum value of $\left(x-r.\dfrac{x}{R}\right)$. The total storage cost per cycle can be calculated by noting the area of the triangle as

$$\tfrac{1}{2}c_2 \frac{x}{r}.\left(x-\frac{rx}{R}\right).$$

The total costs $F(x)$ per unit time are

$$F(x) = \frac{rc_1}{x} + \frac{c_2 x}{2}\left(1-\frac{r}{R}\right).$$

Differentiation gives the optimum batch size as

$$x = \sqrt{\left\{\frac{2rc_1}{c_2(1-r/R)}\right\}}$$

and the interval between batches as

$$\sqrt{\left\{\frac{2c_1}{rc_2(1-r/R)}\right\}}.$$

Substituting the numerical data into these formulae, a production run of 500 items should be made every 10 weeks.

2 The ratio of $\dfrac{c_1}{c_2}.\dfrac{q}{r}$ is $\dfrac{0\cdot5}{100}\left(\dfrac{24}{6}\right) = 0\cdot02$. The least value of y for which

$P_y < 0.02$ is $y = 13$. As the average demand is 6 units per week and the lead time is 1 week, the safety stocks are 13 units.

3 If x magazines are stocked the expected profit $F(x)$ in shillings is the expected revenue less the cost, i.e.

$$F(x) = \sum_{k < x} 6k \cdot p(k) + 6x \sum_{k \geq x} p(k) - 4x$$

where $p(k)$ is the probability of a demand for k magazines. Table 4.5 gives the probabilities for individual demands. The discrete derivative is

$$F(x+1) - F(x) = \sum_{k < x+1} 6k \cdot p(k) + 6(x+1) \sum_{k \geq x+1} p(k) - 4(x+1)$$
$$- \sum_{k < x} 6k \cdot p(k) - 6x \sum_{k \geq x} p(k) + 4x$$
$$= 6 \sum_{k \geq x+1} p(k) - 4.$$

This is positive when

$$\sum_{k \geq x+1} p(k) > \tfrac{2}{3}.$$

Therefore starting from $x = 20$ we want the first value of x for which this inequality does not hold. The optimum value for x is 24. Therefore the optimum number of magazines to stock is 24.

4 Total costs consist of the set-up costs, stock holding cost, and shortage costs.

The set-up costs per unit time $= \dfrac{c_1}{t}$. The average stockholding costs depend on whether demand exceeds n or not. Denoting the demand in a cycle by x, if $x \leq n$, stockholding costs consist of $(n-x)$ units held over the full cycle t and the remainder for a proportional time. Therefore in this case the storage costs per unit time are

$$\frac{c_2}{t} \left[(n-x)t + \tfrac{1}{2}xt \right] = c_2(n - \tfrac{1}{2}x).$$

If $x > n$, the stock is assumed to dwindle to zero over the period $\dfrac{n}{x} \cdot t$ and therefore the storage costs are

$$\frac{c_2}{t} \cdot \tfrac{1}{2}n \cdot \frac{nt}{x} = \frac{c_2 n^2}{2x}.$$

The shortage costs similarly depend on the value of demand.

If $x \leq n$, penalty costs are zero.

If $x > n$ penalty costs $= \dfrac{c_3}{t}(x - n)$.

By taking the expected values the objective function for the total expected cost is

$$F(t, n) = \frac{c_1}{t} + \sum_{x=0}^{n} p(x) \cdot c_2(n - \tfrac{1}{2}x) + \sum_{x=n+1}^{\infty} \left\{ p(x) \cdot \frac{c_2 n^2}{2x} + \frac{c_3}{t}(x - n) \right\}$$

which is to be minimized with respect to t and n.

5 The problem variables are the quantities to purchase x_i and to sell y_i at the start and end of each period $i = 1, \ldots, 5$. The objective function to be maximized is the sum of

$$20y_1 + 35y_2 + 30y_3 + 25y_4 + 50y_5$$
$$- 20x_1 - 22x_2 - 18x_3 - 25x_4 - 40x_5$$
$$- 5(100 + x_1) - 5(100 + x_1 + x_2 - y_1) - \ldots - 5(100 + x_1 + x_2 + \ldots x_5$$
$$- y_1 - y_2 - y_3 - y_4).$$

The storage restriction for period i is

$$100 + (x_1 - y_1) + (x_2 - y_2) + \ldots + x_i \leq 200, \quad i = 1, \ldots, 5.$$

The sales cannot exceed the total stock in any period i,

$$y_i \leq 100 + (x_1 - y_1) + (x_2 - y_2) + \ldots + x_i, \, i = 1, \ldots, 5.$$
$$y_1 \leq 160, \, y_2 \leq 180, \, y_3 \leq 140, \, y_4 \leq 120, \, y_5 \leq 170.$$

The quantities bought must be positive,

$$x_i \geq 0.$$

This completes the linear programming formulation.

6 Let x_{1j}, x_{2j}, x_{3j}, be the number of assemblies to be produced in regular time, overtime and contracted out in the jth month. The quantity overdue in the jth month s_j is the total demanded less the total produced up to and including the jth month, i.e. for successive months:

$$s_1 = 210 - x_{11} - x_{21} - x_{31}$$
$$s_2 = 350 - x_{11} - x_{21} - x_{31} - x_{12} - x_{22} - x_{32}$$
$$s_3 = 530 - x_{11} - x_{21} - x_{31} - x_{12} - x_{22} - x_{32} - x_{13} - x_{23} - x_{33}.$$

The problem variables are x_{ij} and s_j. The total costs consist of the production costs and penalty costs:

$$20 \sum_{j=1}^{4} x_{1j} + 25 \sum_{j=1}^{4} x_{2j} + 30 \sum_{j=1}^{4} x_{3j} + 10(s_1 + s_2 + s_3).$$

The total production must equal the total requirements:

$$\sum_{i=1}^{3} \sum_{j=1}^{4} x_{ij} = 690.$$

The capacity limits on regular and overtime production must be satisfied

$$x_{1j} \leq 150, \quad j = 1, \ldots, 4,$$
$$x_{2j} \leq 30, \quad j = 1, \ldots, 4.$$

Also all the variables x_{ij} and s_j must be positive.

7 The method for solving this problem is described in Example 4.6. Using the same notation the forms of the functions

$g_k(x_k - r_k)$ and $h_k(x_k - x_{k-1})$ are

$$g_k(x_k - r_k) = (x_k - r_k)^2$$
$$h_k(x_k - x_{k-1}) = (x_k - x_{k-1})^2.$$

If y denotes the production level in period $(k-1)$, the recurrence relation for period k is

$$f_k(y) = \min_{x_k \geq r_k} \{(x_k - r_k)^2 + (x_k - y)^2 + f_{k+1}(x_k)\}$$

and $f_5(y) = 0$ for all values of y.

Taking intervals of 5 units we consider the possible values for the production levels y in period 3 to be 90, 95, 100, 105, 110, 115, 120 and record the values of $f_4(y)$ and the corresponding production levels $\bar{x}_4$ as shown in Table A4.1. This table is used to calculate $f_3(y)$ and $\bar{x}_3$ values assuming that y is

Table A4.1

y	90	95	100	105	110	115	120
$f_4(y)$	900	625	400	225	100	25	0
$\bar{x}_4$	120	120	120	120	120	120	120

100, 105, 110, 115, 120, 125, 130 as in Table A4.2. For example $f_3(100)$ is

Table A4.2

y	100	105	110	115	120	125	130
$f_3(y)$	475	450	475	525	600	725	875
$\bar{x}_3$	105	105	105	110	110	110	115

calculated as

$$f_3(100) = \min_{x_3 \geq 90} \{(x_3 - 90)^2 + (x_3 - 100)^2 + f_4(x_3)\}$$

$$= \min \{1000, 675, 500, 475, 600, \ldots\} = 475 \text{ for } x_3 = 105.$$

For the first month we will assume from the data that the production level is 130 and thus we determine $f_2(130)$ as

$$f_2(130) = \min_{x_2 \geq 100} \{(x_2 - 100)^2 + (x_2 - 130)^2 + f_3(x_2)\}$$

$$= 975 \text{ for } \bar{x}_2 = 110.$$

As the production level in month 1 is 130 and in the preceding month it is 120, the production change costs for the first month are 100. Therefore the total costs are $100+975 = 1075$.

The optimal policy is to set the production level for month 1 at 130. This requires that the production level in month 2 should be 110. For a production level of 110 in month 2, Table A4.2 gives the optimal production level in month 3 as 105. Finally with the production level in month 3, at 105, Table A4.1 gives the optimal production level in month 4 as 120.

8 Let x_1, x_2, x_3, x_4, x_5 be the number of transporting, lifting, storing, carrying and hoisting units to be made per day. Then the profit to be maximized is:

$$243x_1 + 616x_2 + 893x_3 + 1367x_4 + 63x_5.$$

The constraints require that

$$45x_1 + 50x_2 + 50x_3 + 175x_4 + 1\cdot5x_5 \leqq \quad 438$$

$$35x_1 + 30x_2 + 30x_3 + 100x_4 + \quad 13x_5 \leqq 1043$$

$$12x_1 + 14x_2 + 15x_3 + \quad 75x_4 + \quad 2x_5 \leqq \quad 633$$

$$x_i \geqq 0, \quad \text{for } i = 1, ..., 5.$$

This is a standard linear programming problem.

9 The problem variables are the quantities of each of the two components produced at each of the factories. Let $x_{ij}(i = 1, 2: j = 1, 4)$ denote the quantity of product i produced at factory j.

The objective is to maximize the weekly output of the finished article. Since the finished article consists of three units of the first component combined with four units of the second the number of finished articles is

$$\min\left(\tfrac{1}{3} \sum_{j=1}^{4} x_{1j}, \tfrac{1}{4} \sum_{j=1}^{4} x_{2j}\right).$$

As there is no advantage in a surplus of either proportion, we can constrain them to be the same, and maximize either, say,

$$\tfrac{1}{3} \sum_{j=1}^{4} x_{1j}.$$

The constraints must now be determined. The first constraint is to ensure that the products are produced in the correct ratio, i.e.

$$\tfrac{1}{3} \sum_{j=1}^{4} x_{1j} = \tfrac{1}{4} \sum_{j=1}^{4} x_{2j}$$

or

$$4 \sum_{j=1}^{4} x_{1j} = 3 \sum_{j=1}^{4} x_{2j}.$$

Secondly the production in each factory is limited by the production rates:
i.e.

$$7x_{11} + 8x_{21} \leq 80$$

$$6x_{12} + 7x_{22} \leq 80$$

$$4x_{13} + 6x_{23} \leq 80$$

$$x_{14} + 5x_{24} \leq 80$$

This optimization problem is now expressed as a standard linear programming problem.

10 Let x_i denote the quantity of item i which is accepted. Then the expected return from the stocking of item i is

$$f_i(x_i) = \sum_{j=0}^{x_i} v_i \cdot j \cdot p_i(j).$$

The total expected return is:

$$F(x_1, x_2, \ldots, x_N) = \sum_{i=1}^{N} f_i(x_i)$$

and this is to be maximized subject to the restriction

$$\sum_{i=1}^{N} x_i w_i \leq C.$$

This problem can be solved by using dynamic programming along the lines described in the last section of Chapter 7.

11 When the lead time itself is subject to statistical variation the problem of determining the safety stock level is more difficult. We can still use average characteristics to determine the typical cycle time and the average batch size. But we now have the problem of requiring additional safety stocks in case the batch arrives late. Using the same notation as was adopted for Example 4.3 we will assume that the distribution of the lead time l is given by $p'(l)$ and that the average lead time is L where

$$L = \sum_{l=0}^{\infty} l \cdot p'(l).$$

The average demand during a lead time is therefore rL and we wish to determine the appropriate safety stock x. The average stockholding costs with safety stocks x will be

$$c_1 x \cdot q/r$$

as before. If the lead time is in fact l, the expected shortage costs will be

$$c_2 \sum_{j=1}^{\infty} j \cdot p(x + rl + j)$$

where $p(x)$ is the probability of x units being demanded per unit time.

However, as l is itself a random variable the expected shortage costs will be

$$c_2 \sum_{j=1}^{\infty} j \cdot \sum_{l=0}^{\infty} p(x+rl+j) \cdot p'(l).$$

The discrete optimization technique which was used in Example 4.3 can now be employed to minimize the revised total costs function.

12 The dynamic programming formulation of the warehousing problem with committed demands is similar to the production smoothing formulation discussed as Example 4.6. As the demands are committed, it is only the cost data which needs to be considered; we therefore minimize costs.

Let $f_k(y)$ be the minimum possible costs over the periods k, $k+1$, ..., N if y is the stock level held in the middle of period $k-1$. Then $f_k(y)$ is made up of the purchase cost in period k plus the storage cost in period k plus the minimum cost over the remaining periods. If x_k is the amount stocked in period k, this gives the recurrence relations

$$f_k(y) = \min \{c_k(x_k - y + r_{k-1}) + cx_k + f_{k+1}(x_k)\}$$

where c is the storage cost and the minimization extends over the range of x_k for which

$$x_k \geq r_k$$
$$x_k \geq y - r_{k-1}$$
$$x_k \leq B.$$

The recurrence relation would be evaluated for a suitable range of values of y and for $k = N$, $N-1$, ..., 1 where $f_{N+1}(x_{N+1}) = 0$. The forwards pass would determine the purchasing policy.

Answers 5

1 Denoting by y_1 the level of property D, it is related to the proportions x_1 and x_2 of metals A and B as:

$$y_1 = d_1 x_1 + d_2 x_2.$$

Using Table 5.3 we have two equations

$$3 = \tfrac{1}{4} d_1 + \tfrac{3}{4} d_2$$
$$6 = \tfrac{3}{4} d_1 + \tfrac{1}{4} d_2.$$

These determine $d_1 = \tfrac{15}{2}$, $d_2 = \tfrac{3}{2}$ and thus the property level y_1 is related to x_1 and x_2 as

$$y_1 = \tfrac{15}{2} x_1 + \tfrac{3}{2} x_2.$$

Similarly the level of property E, say y_2, can be determined as

$$y_2 = e_1 x_1 + e_2 x_2$$

and Table 5.3 gives $e_1 = \frac{5}{2}$, $e_2 = \frac{9}{2}$ providing the relation

$y_2 = \frac{5}{2}x_1 + \frac{9}{2}x_2$.

The customer requires that the level of D should exceed 5 and the level of E should not exceed 4. Therefore if x_1, x_2 and x_3 are the best proportions of metals, we need

$\frac{15}{2}x_1 + \frac{3}{2}x_2 \geq 5$

and

$\frac{5}{2}x_1 + \frac{9}{2}x_2 \leq 4$.

Furthermore as the properties must constitute a unit mix we require

$x_1 + x_2 + x_3 = 1$.

As x_1, x_2, $x_3 \geq 0$, this last constraint may be expressed as

$x_1 + x_2 \leq 1$.

In order to minimize costs we minimize $2x_1 + 3x_2 + x_3$ which using the equality may be expressed as $(1 + x_1 + 2x_2)$. This is a linear programming problem which can be solved graphically to give the optimum mix at the values

$x_1 = \frac{2}{3}$

$x_2 = 0$

$x_3 = \frac{1}{3}$.

2 The basic variables for the reactor are the length and the radius of the cylinder. These quantities will be measured in feet and denoted by l and r. The volume of the reactor V is then

$V = \pi r^2 . l + \frac{4}{3}\pi r^3$.

The operating profit $f_1(r, l)$ is then expressed as the product of the number of hours of operation, the flow rate and the profit margin:

$f_1(r, l) = (8 \times 24 \times 365)\frac{V}{2} . \frac{1}{4} = 8760(\pi r^2 l + \frac{4}{3}\pi r^3)$

The cost of construction is broken down into the material and insulation costs, mixer motor costs and foundation costs.
The surface area of the reactor is

$A = 2\pi r l + 4\pi r^2$.

The material and insulation costs $f_2(r, l)$ are

$f_2(r, l) = A(\frac{1}{4} . 10 + \frac{3}{4} . 5) = 6 \cdot 25(2\pi r l + 4\pi r^2)$.

The mixer motor costs $f_3(r, l)$ are proportional to the square root of the volume as

$f_3(r, l) = 50 . \sqrt{V} = 50(\pi r^2 l + \frac{4}{3}\pi r^3)^{\frac{1}{2}}$.

Finally the foundation costs $f_4(r, l)$ are £500 plus £100 for every foot of radius, i.e.

$$f_4(r, l) = 500 + 100r.$$

The total profit $F(r, l)$ is the difference between the operating profit and the costs:

$$F(r, l) = f_1(r, l) - f_2(r, l) - f_3(r, l) - f_4(r, l)$$
$$= 8760(\pi r 2l + \tfrac{4}{3}\pi r^3) - 6\cdot25(2\pi rl + 4\pi r^2) - 50(\pi r^2 l + \tfrac{4}{3}\pi r^3)^{\frac{1}{2}} - 100r - 500.$$

This is to be maximized subject to the constraints that the volume is between 500 and 1000 cu.ft, i.e.

$$500 \leq \pi r^2 l + \tfrac{4}{3}\pi r^3 \leq 1000$$

and the length is between 6 and 12 feet

$$6 \leq l \leq 12.$$

3 Suppose the standard range for the instrument is $(0, t)$. Then the total costs will be the fraction of demand which uses the standard design multiplied by $(c_0 + c_1 t)$ plus the expected costs for supplying non-standard designs. The distribution of demand is as illustrated in Fig. A5.1. The fraction of demand

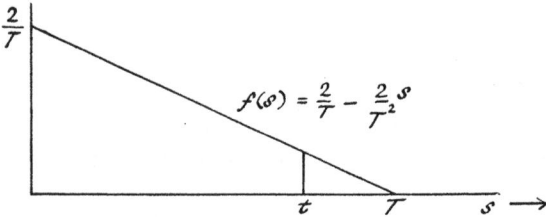

Fig. A5.1

for standard designs is

$$\int_0^t \left(\frac{2}{T} - \frac{2}{T^2} s\right) ds = \frac{2t}{T} - \frac{t^2}{T^2}.$$

Thus the costs for the standard design are

$$\left(\frac{2t}{T} - \frac{t^2}{T^2}\right)(c_0 + c_1 t).$$

The costs for the special designs are

$$\int_t^T \left(\frac{2}{T} - \frac{2}{T^2} s\right)(c_0' + c_1 s) ds = \frac{2}{T}(T - t)c_0' + \left(\frac{c_1}{T} - \frac{c_0'}{T^2}\right)(T^2 - t^2) - \frac{2c_1}{3T^2}(T^3 - t^3).$$

The sum of these costs is a cubic function in t say

$$F(t) = a_0 + a_1 t + a_2 t^2 + a_3 t^3.$$

This can be differentiated to give a quadratic function in t. When it is equated to zero, it will give two roots, one of which will correspond to the value of t for which the costs are minimized.

4 Let p_j denote the machine occupying position j in the factory. Then the allocation of machines to positions is given by the permutation

$$[P] = [p_1, p_2, ..., p_N].$$

The work movement between positions i and j is

$$d(i, j) \cdot v(p_i, p_j).$$

The total work movement to be minimized is

$$F[P] = \sum_{i=1}^{N-1} \sum_{j=i+1}^{N} d(i, j) \cdot v(p_i, p_j).$$

The local optimum can be defined in terms of cyclic interchanges in which we interchange machines in positions i and j or interchange machines in positions i, j and k cyclicly. The disadvantage of applying shift exchanges here is that all the machines' positions will be altered each time a single shift is made, and this could disturb some established good relationships.

5 The network can be expressed as the permutation of elements along the element line. If p_i is the element in position i the permutation is

$$[P] = [p_1, p_2, ..., p_N].$$

Also the number of crossings is obtained similarly to the formula in Example 5.7 as

$$F[P] = \tfrac{1}{2} \sum_{i=1}^{N-1} \sum_{j=i+1}^{N} A(i, j) \left[\sum_{k=1}^{i-1} \sum_{l=i+1}^{j-1} A(k, l) + \sum_{k=i+1}^{j-1} \sum_{l=j+1}^{N} A(k, l) \right]$$

where $A(i, j) = 1$ if (p_i, p_j) are a connected pair of elements,
            $= 0$ otherwise.

Using the data in Fig. 5.9 a network can be determined by the same scheme as was used for the permutation procedure in Example 5.7. We start with the element having most connections, number 2, and then introduce the elements most connected to the already placed elements, placing them in their best positions. This gives a network with 2 crossings as shown in Fig. A5.2.

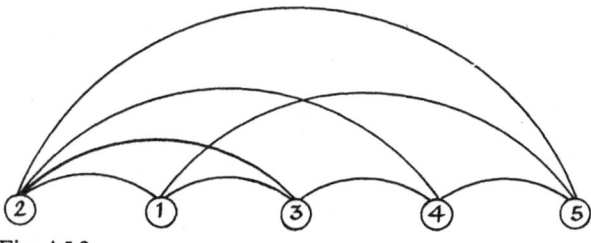

Fig. A5.2

Using this as the base permutation $P^{(0)}$, we can draw up the permutations one adjacent exchange away to obtain their crossings value as given below.

	Number of crossings
Permutation	
$[P^{(0)}] = [2, 1, 3, 4, 5]$	2
$[P^{(1)}] = [1, 2, 3, 4, 5]$	2
$[P^{(2)}] = [2, 3, 1, 4, 5]$	3
$[P^{(3)}] = [2, 1, 4, 3, 5]$	3
$[P^{(4)}] = [2, 1, 3, 5, 4]$	3

Hence the initial permutation $[P^{(0)}]$ is optimal with respect to adjacent exchanges.

Answers 6

1 The average annual costs for machine A can be calculated from Table A6.1.

Table A6.1

Machine A

Replace at end of year	Total running costs	Total capital costs	Total costs	Average annual costs
1	800	3000	3800	3800
2	1600	3000	4600	2300
3	2400	3000	5400	1800
4	3400	3000	6400	1600
5	4600	3000	7600	1520
6	6000	3000	9000	1500
7	7600	3000	10600	1514

We therefore replace machine A every 6 years and the average annual cost is £1500.

The costs for machine B are calculated in Table A6.2.

Table A6.2

Machine B

Replace at end of year	Total running costs	Total capital costs	Total costs	Average annual costs
1	1200	1500	2700	2700
2	2400	1500	3900	1950
3	3600	1500	5100	1700
4	4800	1500	6300	1575
5	6200	1500	7700	1540
6	7800	1500	9300	1550

Machine B is therefore replaced every 5 years at average annual cost of £1540. Machine A offers the cheaper average annual cost and would be preferred.

2 If the machine is replaced every n years the average operating cost over the n years is

$$\frac{0+u+2u+...+(n-1)u}{n} = \frac{1}{n} = \frac{n(n-1)u}{2} = \frac{(n-1)u}{2}.$$

The average capital cost over n years is

$$\frac{C-s}{n}.$$

The total cost $F(n)$ is measured as

$$F(n) = \frac{(n-1)u}{2} + \frac{C-s}{n}.$$

By differentiating we obtain the minimum value of n as

$$n = \sqrt{\left\{\frac{2(C-s)}{u}\right\}}.$$

3 Using Table 6.8, the value of $f_5(1)$ is the optimal return over years 5 to 10. Therefore the optimal return will be 390, and the optimal policy will be to keep the machine in years 5, 6, 7, replace it in year 8 and keep the new machine in years 9 and 10.

4 Define $f_n(t)$ as the minimum cost if the owner has a car of age t at the beginning of year n. Let $u(t)$ denote the running cost of a car of age t let $s(t)$ denote the trade-in price of a car of age t years, and let $C(t)$ denote the buying price of a car of age t. At the start of year n he can either trade in the existing car of age t and purchase a car of age x (where $x = 0$ denotes a new car) and then make a further decision a year later with a car of age $(x+1)$ or else he can keep the current car which will be of age $(t+1)$ at the end of the year. As all quantities are measured as costs, he wishes to choose the policy which minimizes the total costs. Therefore the dynamic programming recurrence relations are obtained as

$$f_n(t) = \min \begin{cases} \text{Replace with a car of age } x \text{ at cost} \\ \min_{0 \le x \le 5} (C(x)+u(x)-s(t)+f_{n+1}(x+1)). \\ \text{Keep the current car at cost} \\ u(t)+f_{n+1}(t+1). \end{cases}$$

The recurrence relations must be evaluated over a six-year period and also for $1 \le t \le 6$ and we can set $f_7(t) = 0$ for all values of t. The policy of keeping will be denoted by (K) and the replacement by a car of age x by

$(R(x))$. First we calculate $f_6(t)$, then $f_5(t)$, and so on. The results are shown in Table A6.3.

Table A6.3

t	$f_6(t)$	Policy	$f_5(t)$	Policy	$f_4(t)$	Policy	$f_3(t)$	Policy	$f_2(t)$	Policy	$f_1(t)$	Policy
1	0	$R(3)$	300	$R(2)$	625	$R(2)$	925	$R(2)$	1250	$R(2)$	1575	$R(2)$
2	200	$R(3)$	450	K	775	K	1075	K	1400	K	1700	K
3	250	K	575	$R(2)$	875	K	1200	$R(2)$	1500	K	1825	K
4	325	$R(3)$	625	$R(2)$	950	$R(2)$	1250	$R(2)$	1575	$R(2)$	1900	$R(2)$
5	370	$R(3)$	670	$R(2)$	995	$R(2)$	1295	$R(2)$	1620	$R(2)$	1945	$R(2)$
6	390	$R(3)$	690	$R(2)$	1015	$R(2)$	1315	$R(2)$	1640	$R(2)$	1965	$R(2)$

For example, $f_6(3) = 250$, as the cost of keeping a three-year-old car in year 6 is the running cost £250, whereas the least replacement cost is to take on a three-year-old car with one year's running costs at $250 + 150 = 400$ less the trade-in price £125 giving a net cost of £275. Once $f_6(t)$ has been calculated it can be assessed that the minimum value of

$$\min_{0 \le x \le 5} \ (C(x) + u(x) + f_6(x+1)) = 250 + 250 + 200 = 700$$

occurs for $x = 2$. This can be used to determine the $f_2(t)$ values as it is independent of t. For example, the cost of replacing a three-year-old car is £700 less the trade-in price £125 giving a net cost of £575. If a three-year-old car is kept the cost is

$$250 + f_6(4) = 250 + 325 = 575.$$

In this case either policy is equally good.

The table indicates the optimal policies. For instance, if we start with a car of age 1 in year 1 it should be replaced with a car of age 3 in year 1, which in turn is replaced with a two-year-old car in year 2, which is replaced by another two-year-old car in year 3, kept in year 4, replaced in year 5 by another two-year-old car and kept in year 6. The whole policy costs £1575. The optimal policies for various ages of initial car are given in Table A6.4.

Table A6.4

Age of car in year 1	Decision in year						Total cost
	1	2	3	4	5	6	
1	$R(3)$	$R(2)$	$R(2)$	K	$R(2)$	K	1575
2	K	K	$R(2)$	K	$R(2)$	K	1700
3	K	$R(2)$	$R(2)$	K	$R(2)$	K	1825
4	$R(3)$	$R(2)$	$R(2)$	K	$R(2)$	K	1900
5	$R(3)$	$R(2)$	$R(2)$	K	$R(2)$	K	1945
6	$R(3)$	$R(2)$	$R(2)$	K	$R(2)$	K	1965

5 Let x_1 and x_2 denote the quantities of equipment of types 1 and 2 to buy. Then the capital costs are $c_1x_1+c_2x_2$. It is required that the capacity should equal the maximum demand, i.e.

$$x_1+x_2 = A.$$

The total running costs in the peak period will be

$$a(r_1x_1+r_2x_2).$$

The total running costs in the off-peak period will depend on the proportion of the different capacities which operate in the off-peak. Let y_1 and y_2 denote the amount of equipment 1 and 2 operating in the off-peak period. Then we require

$$y_1+y_2 = B$$

and the running costs in the off-peak period are

$$b(r_1y_1+r_2y_2).$$

Therefore the aim is to choose x_1, x_2, y_1, y_2 to minimize the total costs

$$(c_1+ar_1)x_1+(c_2+ar_2)x_2+br_1y_1+br_2y_2$$

subject to

$$x_1+x_2 = A$$
$$y_1+y_2 = B$$
$$y_1 \leqq x_1$$
$$y_2 \leqq x_2$$
$$x_1, x_2, y_1, y_2 \geqq 0.$$

This linear programming problem is known as the peak demand problem.

6 The decision variables are the quantities x_{ij} to manufacture at site i for destination j in each year and the zero-one variables y_i to decide whether to set up a plant at site i.

The total costs over the five years consist of the set-up costs, production costs and transportation costs over the five years

$$50{,}000y_1+30{,}000y_2+65{,}000y_3$$
$$+5[400(x_{11}+x_{12}+x_{13}+x_{14}+x_{15}+x_{16})$$
$$+500(x_{21}+x_{22}+x_{23}+x_{24}+x_{25}+x_{26})$$
$$+300(x_{31}+x_{32}+x_{33}+x_{34}+x_{35}+x_{36})$$
$$+10\{27x_{11}+17x_{12}+35x_{13}+42x_{14}+16x_{15}+12x_{16}$$
$$+41x_{21}+36x_{22}+18x_{23}+10x_{24}+56x_{25}+28x_{26}$$
$$+22x_{31}+49x_{32}+5x_{33}+25x_{34}+29x_{35}+46x_{36}\}].$$

The constraints require that the production capabilities should not be exceeded, i.e.

$$\sum_{j=1}^{6} x_{1j} \leqq 35,000$$

$$\sum_{j=1}^{6} x_{2j} \leqq 35,000$$

$$\sum_{j=1}^{6} x_{3j} \leqq 25,000;$$

that the demands should be satisfied

$$\sum_{i=1}^{3} x_{i1} = 15$$

$$\sum_{i=1}^{3} x_{i2} = 7$$

$$\sum_{i=1}^{3} x_{i3} = 10$$

$$\sum_{i=1}^{3} x_{i4} = 8$$

$$\sum_{i=1}^{3} x_{i5} = 6$$

$$\sum_{i=1}^{3} x_{i6} = 12.$$

No production can take place unless a plant has been established, therefore

if $y_i = 0, \quad x_{ij} = 0;$

if $y_i = 1, \quad x_{ij} \geqq 0.$

7 Let x_k be the total quantity of plant of type k installed in the year. Then if $h_k(x_1, x_2, ..., x_M)$ denotes the number of hours of operation of plants of type k, the total costs are

$$F(x_1, x_2, ..., x_M) = \sum_{k=1}^{M} (a_k + r_k h_k(x_1, x_2, ..., x_M)x_k.)$$

The $h_k(x_1, x_2, ..., x_M)$ can be determined by arranging the plants in a merit order. The total amount of plant operating with cheaper running costs than new installation of type k is approximated as

$$x_1 + x_2 + ... + x_{k-1} + \frac{x_k}{2}.$$

The corresponding number of hours of operation is then

$$h_k(x_1, x_2, \ldots, x_M) = H\left(x_1 + x_2 + \ldots + x_{k-1} + \frac{x_k}{2}\right).$$

The objective function then becomes

$$\sum_{k=1}^{M}\left[a_k + r_k \cdot H\left(x_1 + x_2 + \ldots + x_{k-1} + \frac{x_k}{2}\right)\right]x_k.$$

The constraints require simply that

$x_k \geqq G_k$, for $k = 1, \ldots, M$

and

$$x_1 + x_2 + \ldots + x_k = D.$$

To solve this problem by dynamic programming, let us define $f_k(y)$ as the least cost of filling a capacity of y megawatts with the first k types of plant. Then if x_k megawatts of plant type k are installed, $f_k(y)$ consists of the costs attributable to these x_k units plus the remaining costs due to plant types $x_1, x_2, \ldots, x_{k-1}$. Thus we may write

$$f_k(y) = \min_{G_k \leqq x_k \leqq y}\left\{x_k\left[a_k + r_k H\left(y - \frac{x_k}{2}\right)\right] + f_{k-1}(y - x_k)\right\}.$$

This recurrence relations must be evaluated for

$G_k \leqq y \leqq D$

and for $k = 1, 2, \ldots, M$, where $f_0(y) = 0$. If the x_k quantities say $\bar{x}_k(y)$ are recorded for which the minimization is obtained for each value of y, the optimal installation pattern will be determined.

8 Let d_i denote the demand level at point i and let (x_i, y_i) denote its co-ordinates. If one depot is to be established, it will be located at the weighted centre of gravity, i.e. at the point $(\bar{x}, \bar{y})$ where

$$\bar{x} = \sum_{i=1}^{6} x_i d_i \bigg/ \sum_{i=1}^{6} d_i = 3 \cdot 1$$

$\bar{y} = 2 \cdot 48$.

The distance from $(\bar{x}, \bar{y})$ to the various points can be determined graphically to give the results shown in Table A6.5.

Table A6.5

Point number	1	2	3	4	5	6
Distance from $(\bar{x}, \bar{y})$	4·0	2·2	2·6	1·0	2·5	5·1

The total number of units—miles per week— is then the sum of the products of the distances and the demand levels, i.e.

110 (4·0)+80 (2·2)+40 (2·6)+140 (1·0)+30 (2·5)+100 (5·1)
= 1445.

The total costs per week are then

$$\frac{1445}{4} + \frac{5000}{52}$$

= 361+96 = £457.

If two depots are established with exclusive markets, their locations are first determined by the heuristic procedure.

The two points most distant are points 1 and 6. These are the first two tentative locations for the first and second groups. We now proceed iteratively.

Iteration 1

Point 5 is the nearest point, and it is added to group 2. The tentative locations are denoted by $(\bar{x}_1, \bar{y}_1)$ for group 1 and by $(\bar{x}_2, \bar{y}_2)$ for group 2.

$(\bar{x}, \bar{y}_1) = (0, 0)$
$(\bar{x}_2, \bar{y}_2) = (5·8, 5·5)$.

Iteration 2

Point 2 is the next nearest point and it is added to group 1. Group 1's new location is determined as

$(\bar{x}_1, \bar{y}_1) = (0·42, 1·3)$

Iteration 3

Point 4 is now added to group 1.

It is now clear from graphical inspection that point 3 will be added to group 1. The final groupings will then be:

Group 1: 1, 2, 3 and 4

Group 2: 5 and 6.

The locations of the depots for the two groups are

Group 1: $(\bar{x}_1, \bar{y}_1) = (2·16, 1·4)$
Group 2: $(\bar{x}_2, \bar{y}_2)$ (5·8, 5·5).

The distances of the points from their respective depots are given in Table A6.6.

Table A6.6

Point Number	1	2	3	4	5	6
Distance from $(\bar{x}_1, \bar{y}_1)$ or $(\bar{x}_2, \bar{y}_2)$	2·6	2·0	2·4	2·0	1·6	0·6

The total number of units-miles per week is then

110 (2·6)+80 (2·0)+40 (2·4)+140 (2·0)+30 (1·6)+100 (0·6) = 930.

The total costs per week for the two depots are then

$$\frac{930}{4} + \frac{10{,}000}{52}$$

= 232+192 = £424.

Therefore it is cheaper to establish two depots than one as the savings in transportation costs more than balances the extra set-up costs incurred.

9 Under conditions of no technological change the operating costs $u(t)$, the returns $r(t)$, the salvage value $s(t)$ for a machine of age t and the cost of a new machine C do not change over time. For dynamic programming purposes define the function $f(t)$ as the discounted present value of all capital and operating costs associated with a machine of age t years employing an optimal replacement policy, where $t = 1, 2, 3, \ldots$ At any time t there are two alternatives: to replace or to keep the machine for a further time period. If the present machine is kept, and it is currently t years old, the present value of the revenue consists of the returns less the running costs of a t-year-old machine plus all future costs of a $(t+1)$-year-old machine:

$r(t)-u(t)+df(t+1)$.

If the machine is replaced, the returns consist of the revenue less the cost on a 1-year-old machine less the capital cost plus the salvage value

$r(1)-u(1)-C+s(t)+df(2)$.

Thus for an optimal policy:

$$f(t) = \min \begin{cases} \text{Keep:} & r(t)-u(t)+df(t+1) \\ \text{Replace:} & r(1)-u(1)-C+s(t)+df(2). \end{cases}$$

In the stable conditions of no technological change, the optimal life span T is constant, and thus

$$f(1) \quad = r(1)-u(1)+df(2)$$
$$f(2) \quad = r(2)-u(2)+df(3)$$
$$\vdots \qquad \qquad \vdots \qquad \vdots$$
$$f(\dot{T}-1) = r(\dot{T}-1)-u(T-1)+df(T)$$
$$f(T) \quad = r(1)-u(1)-C+s(T)+df(1).$$

These equations can be solved for $f(1)$ by successive substitution of $f(T)$ into $f(T-1), f(T-1)$ into $f(T-2)$ and so on yielding the following results:

$$f(T-1) = r(T-1)-u(T-1)+d(r(1)-u(1)-C+s(T)+d\cdot f(1))$$
$$f(T-2) = r(T-2)-u(T-2)+d(r(T-1)-u(T-1))$$
$$+d^2(r(1)-u(1)-C+s(T)+d\cdot f(1))$$
$$f(1) = \sum_{t=1}^{T-1} (r(t)-u(t))d^{t-1}+(r(1)-u(1)-C+s(T)+d\cdot f(1))d^{T-1}.$$

This gives the result for $f(1)$:

$$(1-d^T)f(1) = \sum_{t=1}^{T-1}(r(t)-u(t))d^{t-1}+(r(1)-u(1)-C+s(T))d^{T-1}$$

or

$$f(1) = \left(\sum_{t=1}^{T-1}(r(t)-u(t))d^{t-1}+(r(1)-u(1)-C+s(T))d^{T-1}\right)/(1-d^T).$$

This is identical to the result obtained in Section 6.2.

Answers 7

1 Let x_{ij} be the number of buses to be despatched from G_i to P_j. Then we wish to minimize the total travel time

$$5x_{11}+4x_{12}+3x_{13}+2x_{14}$$
$$+10x_{21}+8x_{22}+4x_{23}+7x_{24}$$
$$+\ 9x_{31}+9x_{32}+8x_{33}+4x_{34}$$

subject to the supply constraints

$$x_{i1}+x_{i2}+x_{i3}+x_{i4}=5 \text{ for } i=1,\,2,\,3$$

Table A7.1

		Demands			
		1	6	2	6
v_j u_i		5	4	0	3
5	0	5 ①	4 — ④	3 ⊡	2 + ③
5	4	10 ⑨	8 ② +	4 ②	7 ①
5	1	9 ⑥	9 ⑤	8 ①	4 ⑤

and the demand constraints

$$x_{11}+x_{21}+x_{31}=1$$
$$x_{12}+x_{22}+x_{32}=6$$
$$x_{13}+x_{23}+x_{33}=2$$
$$x_{14}+x_{24}+x_{34}=6.$$

All $x_{ij} \geq 0$.

This is a standard transportation problem as described in Section 7.2. It can be solved by the method described in Section 7.3. Using the North-West corner rule an initial allocation is obtained as shown in Table A7.1, in which the same symbol conventions have been used as for Table 7.4.

The only unit cost saving for the measure $u_i + v_j - c_{ij}$ occurs on the route (1, 4), one unit is put on to this route on the circuit marked in Table A7.1. The route (2, 4) is dropped from the basis and the new allocation as shown in Table A7.2.

As there are no values of (i, j) for which

$$u_i + v_j - c_{ij}$$

is positive for the non-basic routes this is an optimal allocation.

Table A7.2

		Demands			
		1	6	2	6
	v_j / u_i	5	4	0	2
5	0	5 (1)	4 (3)	3 [0]	2 (1)
Supplies					
5	4	10 [9]	8 (3)	4 (2)	7 [6]
5	2	9 [7]	9 [6]	8 [2]	4 (5)

2 As explained in Section 7.4 on the transhipment problem, the cost matrix will be a 7×7 matrix as shown.

The supplies and demands will need to be increased artificially to allow transhipment to take place. As the total number of units is 50 we can add 30 units to the supplies at each point and 30 units to the demands. Thus if p_i denotes the supply at source i and q_i the demand at destination i they are as shown.

We now have a standard transportation problem with supplies p_i, demands q_j and a cost matrix c_{ij} as given in Table A7.3, and this can be solved by a standard method.

It is interesting to note the way in which the matrix operates. Suppose 5 units are moved from S_3 to S_1. This will restrict the value of x_{11} (the amount shipped at zero cost from S_1 to S_1) to an upper limit of 25 as the sum of the x_{ij} values relating to the first column of the matrix in Table A7.3

I

Table A7.3

		To						
Destination		1	2	3	4	5	6	7
Source		S_1	S_2	S_3	D_1	D_2	D_3	D_4
1	S_1	0	3	2	5	9	6	8
2	S_2	2	0	3	8	4	6	6
3	S_3	2	8	0	5	6	8	7
From 4	D_1	5	8	5	0	5	2	3
5	D_2	9	4	6	5	0	1	8
6	D_3	6	6	8	4	2	0	4
7	D_4	8	6	7	5	3	9	0

must be 30. As the supply at S_1 is 50 a total of 25 units will be distributed to the demand points D_1, D_2, D_3, D_4 whereas, without the possibility of transhipment the quantity available for movement was 20 units.

Table A7.4

i	1	2	3	4	5	6	7
p_i	50	50	40	30	30	30	30
q_i	30	30	30	46	40	40	44

3 When the dynamic programming procedure is applied to the network the results are as shown in Table A7.5.

Table A7.5

Stage	1		2		3		4		5	
Point j	$f_1(j), q_0(j)$		$f_2(j), q_1(j)$		$f_3(j), q_2(j)$		$f_4(j), q_3(j)$		$f_5(j), q_4(j)$	
1	0	—	0	—	0	—	0	—	0	—
2	3	1	3	1	3	1	3	1	3	1
3	6	1	4	4	4	4	4	4	4	4
4	1	1	1	1	1	1	1	1	1	1
5	∞	—	6	4	6	4	6	4	6	4
6	4	1	4	1	4	1	4	1	4	1
7	∞	—	7	2	6	3	6	3	6	3
8	∞	—	11	6	11	6	10	7	10	7

The shortest route has length 10 units and passes through the points 1, 4, 3, 7, 8.

4 Using the heuristic of savings criterion described in Example 7.4.

Step 1. Select link 7–8.

Step 2. Select link 4–5.

Step 3. Select link 4–7 giving route 1, 8, 7, 4, 5, 1 which has length 64 and a total load of $1+4+3+2 = 10$ so that it satisfies both the mileage and capacity limits. All links with 8, 7, 4, or 5 in them can be eliminated.

Step 4. Select link 3–9.

Step 5. Select link 2–9. Not feasible because the weight limit when linked up with 3-9 is exceeded.

Step 6. Select link 2-6.

No further link is feasible.

This gives 3 routes 1, 8, 7, 4, 5, 1 with distance 64 and load 10

1, 2, 6, 1	with distance 35 and load 8
1, 3, 9, 1	with distance 32 and load 8.

5 Let n be the number of vehicles which the firm owns, let x_i be the number of owned vehicles which the firm runs on day i and let y_i be the number of vehicles which the firm hires on day i. Then the objective function to be minimized is

$$F(n, x, y) = cn+v \sum_{i=1}^{N} x_i+h \sum_{i=1}^{N} y_i.$$

The constraints are:
$0 \leq x_i \leq n$
$x_i+y_i \geq d_i$
$y_i \geq 0.$

This problem becomes more difficult, although more realistic, when a number of types of vehicles are considered having different capacities and the loads of the individual distributions must be assessed.

6 First we use Tables 7.21 and 7.22 to calculate the schedule of arrivals at the destinations Q_i and order them as a sequence $i = 1$ to 7 as shown in Table A7.6, noting the number of ships as $m(i)$.

Table A7.6

Schedule i	1	2	3	4	5	6	7
Arrival at	Q_2	Q_1	Q_2	Q_1	Q_3	Q_2	Q_3
Time	11	18	18	21	41	43	54
No. of ships $m(i)$	1	2	1	2	1	1	1

I*

The data of Table 7.23 can be arranged to present the starting sequence as shown in Table A7.7 noting the number of ships required as $n(j)$.

Table A7.7

Schedule j	1	2	3	4	5	6	7
Departure from	P_3	P_1	P_1	P_2	P_3	P_1	P_2
Time	0	8	14	16	32	35	46
No. of ships $n(j)$	1	2	1	2	1	1	1

Now let x_{ij} be the number of ships to be moved from position i in the arrival schedule to position j in the departure schedule. Then we wish to maximize the number of such transfers

$$\sum_{i=1}^{7} \sum_{j=1}^{7} x_{ij}$$

subject to the restrictions that

$x_{ij} \leqq m(i)$

$x_{ij} \leqq n(j)$

x_{ij} integer

and $x_{ij} = 0$ for all positions marked $\times$ in the following 7×7 matrix as these linkages are impossible, the time of travel exceeding the difference between the departure time and the arrival time.

Table A7.8

Departure schedule

×	×	×				
×	×	×	×			
×	×	×	×			
×	×	×	×			
×	×	×	×	×	×	×
×	×	×	×	×	×	×
×	×	×	×	×	×	×

Arrival schedule

It should be noted that if a computer program was being used to solve this problem the $x_{ij} = 0$ positions could be determined automatically.

7 This problem can be formulated as a permutation problem. Let p_i be the route of the jth train to be sequenced through the junction. Then the order is specified by the permutation

$$[P] = [p_1, p_2, \ldots, p_N].$$

The time $T(p_j)$ at which the jth train passes through the junction is

$$T(p_j) = \max \, (a_j, \, T(p_{j-1})+t(p_{j-1}, \, p_j)).$$

Thus the objective function value is

$$F[P] = \max_{1 \,\leqq\, j \,\leqq\, N} \, (T(p_j)-a_j).$$

All permutations are feasible.

To solve the problem we could define the local optimum by a subset of single shifts limiting the movement to a neighbouring range of positions in the permutation. The initial permutation could be constructed by ordering the trains by increasing a_j values.

8 Suppose we have N customers located at points 1, 2, ..., N with demand quantity $q(i)$ required at point i and distance $d(i, j)$ between points i and j. The vehicle is restricted to a mileage limit of M and a capacity of limit of Q.

The problem of representing the series of routes need not explicitly include the depot. The sequence of customer points visited on the routes can be represented by a single permutation

$$[P] = [p_1, p_2, ..., p_N].$$

If the depot is denoted by point 0 the first route is determined as the sequence

$$0-p_1-p_2-...p_k-0$$

where k is determined such that

$$d(0, p_1)+d(p_1, p_2)+...d(p_{k-1}, p_k)+d(p_k, 0) \leq M$$

and

$$q(p_1)+q(p_2)+... q(p_k) \leq Q$$

but one of these constraints is violated if p_{k+1} is added on to the end of the route. The next route is then worked out starting at point p_{k+1}, and so on until all points have been included in a route. We wish to determine the permutation which minimizes the number of routes and, for them, minimizes the total distance covered. This formulation can be used to determine a permutation which is locally optimal permutation with respect to a chosen set of exchanges.

9 The dynamic programming procedure states that the iterations can be stopped at iteration n, when

$$f_n(j) = f_{n-1}(j) \text{ for } j = 1, ..., N.$$

We need to be sure that $f_n(j)$ could not be reduced at a subsequent iteration. $f_n(j)$ is calculated by the relation

$$f_n(j) = \min_{1 \,\leqq\, i \,\leqq\, N} \, \{f_{n-1}(i)+d(i, j)\}.$$

As $f_n(j) = f_{n-1}(j)$ for all j the same quantities will be involved in the calculation of $f_{n+1}(j)$. Therefore

$$\begin{aligned}
_{+1}(j) &= \min_{1 \leq i \leq N} \{f_n(i) + d(i, j)\} \\
&= \min_{1 \leq i \leq N} \{f_{n-1}(i) + d(i, j)\} \\
&= f_n(j).
\end{aligned}$$

This applies for all j and shows that no further change will occur.

Answers 8

1 Let £x_i be the amount invested in security type i. Then we wish to maximize the total yield:

$$3 \cdot 5x_1 + 2 \cdot 5x_2 + 3x_3 + 4 \cdot 5x_4 + 5x_5 + 4x_6$$

subject to the fund limit:

$$\sum_{i=1}^{6} x_i \leq 100{,}000$$

and the constraints

$$100(x_1 + x_2) \geq 35(x_1 + x_2 + x_3 + x_4 + x_5 + x_6)$$
$$100(x_3 + x_4) \leq 40(x_1 + x_2 + x_3 + x_4 + x_5 + x_6)$$
$$100(x_5 + x_6) \leq 40(x_1 + x_2 + x_3 + x_4 + x_5 + x_6)$$

This is a straightforward linear programming problem.

2 Let x_1, x_2, x_3 be the quantities of securities 1, 2 and 3 which the investor purchases. He wishes to maximize his rate of return

$$6x_1 + 4x_2 + 8x_3$$

subject to not exceeding up his fund:

$$10x_1 + 5x_2 + 5x_3 \leq 150$$

and ensuring that the variance on the return is less than 1000

$$1 \cdot 5x_1^2 + x_2^2 + 2 \cdot 5x_3^2 \leq 1000$$

where $x_1, x_2, x_3 \geq 0$.

The third security has the largest return. If all £150 are spent on this security he will obtain the solution

$$x_1 = x_2 = 0$$

as $x_3 = 20$,

$x_3 \leq 30$ by the first constraint
$x_3 \leq 20$ by the second constraint.

Therefore

the return is 160.

However if we reduce the purchase of x_3 to 19 units and increase x_1 and x_2 to 1 and 1 respectively the two constraints are still satisfied and the total return is increased by 2 units. Thus it will not be the best policy to purchase as much as possible of the third security.

3 The method of dynamic programming can be used to solve this problem using the recurrence relations associated with the standard allocation of a single resource. The table below shows the results in the usual tabular form. The entry in the jth row of the kth column shows the maximum response which can be obtained by allocating j hundred pounds to the first k media and the entry in parentheses shows the amount to be allocated to the kth medium. The maximum response value is 13 and this is obtained by assigning £300 to medium 4 and £100 to each of media 1, 2 and 3.

Table A8.1

		1	2	3	4
	0	0 (0)	0 (0)	0 (0)	0 (0)
	1	4 (1)	4 (0)	4 (0)	4 (0)
Total	2	4 (2)	5 (1)	6 (1)	7 (1)
allocation	3	6 (3)	6 (0)	7 (1)	9 (1)
£100	4	6 (4)	7 (1)	8 (1)	10 (1)
	5	7 (5)	8 (2)	9 (1)	12 (3)
	6	7 (6)	11 (5)	11 (0)	13 (3)

4 The data gives a value for the parameters in the formulae of Section 8.7 as

$$a = 1000$$

$$p = 0\cdot1$$

$$x = \tfrac{1}{4}.$$

If the advertisement takes place on day 1, the level of awareness $e(d)$ on day d is given by the formula

$$e(d) = \sum_{t=1}^{d} 1000(0\cdot6)^{d-t}(1-0\cdot1)(0\cdot1)^{t-1}\cdot\tfrac{1}{2}.$$

This gives

$e(1)\ \ = 450$

$e(2)\ \ = 270+45 = 315$

$e(3)\ \ = 162+27+4 = 193$

$e(4)\ \ = 97+16+2+0 = 115$

$e(5)\ \ = 58+10+1+0 = 69$

$e(6)\ \ = (0\cdot6)\,68 = 41$

$e(7)\ \ = (0\cdot6)\,41 = 25$

$e(8)\ \ = 14$

$e(9)\ \ = 8$

$e(10) = 5$

$e(11) = 3$

$e(12) = 2$

$e(13) = 1.$

Therefore after 14 days the awareness of the advertisement will have dwindled to zero.

Index